# Student's Solutions Manual to Accompany Streeter/Hutchison/Hoelzle Basic Mathematical Skills with Geometry Form A

## Third Edition

**John R. Martin**
*Tarrant County Junior College*

**McGraw-Hill, Inc.**
New York St. Louis San Francisco Auckland Bogotá
Caracas Lisbon London Madrid Mexico City Milan
Montreal New Delhi San Juan Singapore
Sydney Tokyo Toronto

STUDENT'S SOLUTIONS MANUAL TO ACCOMPANY
STREETER/HUTCHISON/HOELZLE
BASIC MATHEMATICAL SKILLS WITH GEOMETRY, FORM A



3 4 5 6 7 8 9 0 MAL MAL 9 0 9 8 7 6 5 4

ISBN 0-07-063012-7

The editor was Karen M. Minette;
the production supervisor was Al Rihner.
Malloy Lithographing, Inc., was printer and binder.

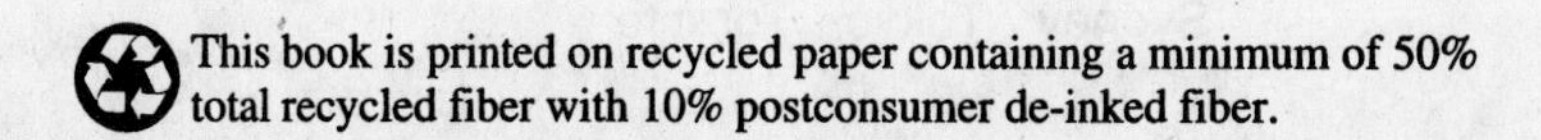
This book is printed on recycled paper containing a minimum of 50% total recycled fiber with 10% postconsumer de-inked fiber.

# TABLE OF CONTENTS

| | | Page |
|---|---|---|
| 1. | Addition and Subtraction of Whole Numbers | 1 |
| 2. | Multiplication of Whole Numbers | 9 |
| 3. | Division of Whole Numbers | 15 |
| 4. | Factors and Multiples | 23 |
| 5. | An Introduction to Fractions | 25 |
| 6. | Multiplication and Division of Fractions | 29 |
| 7. | Addition and Subtraction of Fractions | 33 |
| 8. | Addition, Subtraction, and Multiplication of Decimals | 43 |
| 9. | Division of Decimals | 47 |
| 10. | Ratios and Proportion | 55 |
| 11. | Percent | 61 |
| 12. | The English System of Measurement | 67 |
| 13. | The Metric System of Measurement | 73 |
| 14. | Geometry | 75 |
| 15. | Statistics | 79 |
| 16. | The Integers | 83 |
| 17. | Algebraic Expressions and Equations | 87 |

# CHAPTER 1

# ADDITION AND SUBTRACTION OF WHOLE NUMBERS

## EXERCISES 1.1

1. $352 = 3 \cdot 100 + 5 \cdot 10 + 2 \cdot 1$
3. $5073 = 5 \cdot 1000 + 7 \cdot 10 + 3 \cdot 1$
5. 3 in 327 is in hundreds place
7. 6 in 56,489 is in thousands place
9. 7 in 27,243,012 is in millions place
11. 2 in 523,010,000 is in ten millions place
13. 3456 : Three thousand, four hundred fifty–six
15. 200,304 : Two hundred thousand, three hundred four
17. Two hundred fifty–three thousand, four hundred eighty–three : 253,483
19. Five hundred two million, seventy–eight thousand : 502,078,000
21. Eight : 8

## EXERCISES 1.2

1. In 5 + 4 = 9, 5 is the addend, 4 is the addend, and 9 is the sum.

3. $\begin{array}{r} 4 \\ +\ 3 \\ \hline 7 \end{array}$ 5. $\begin{array}{r} 9 \\ +\ 5 \\ \hline 14 \end{array}$ 7. $\begin{array}{r} 7 \\ +\ 3 \\ \hline 10 \end{array}$ 9. $\begin{array}{r} 4 \\ +\ 4 \\ \hline 8 \end{array}$ 11. $\begin{array}{r} 8 \\ +\ 6 \\ \hline 14 \end{array}$

13. $\begin{array}{r} 7 \\ +\ 4 \\ \hline 11 \end{array}$ 15. $\begin{array}{r} 6 \\ +\ 6 \\ \hline 12 \end{array}$ 17. $\begin{array}{r} 9 \\ +\ 1 \\ \hline 10 \end{array}$ 19. $\begin{array}{r} 8 \\ +\ 7 \\ \hline 15 \end{array}$ 21. $\begin{array}{r} 5 \\ +\ 4 \\ \hline 9 \end{array}$

23. $\begin{array}{r} 3 \\ +\ 5 \\ \hline 8 \end{array}$ 25. $\begin{array}{r} 5 \\ +\ 5 \\ \hline 10 \end{array}$ 27. $\begin{array}{r} 7 \\ +\ 0 \\ \hline 7 \end{array}$ 29. $\begin{array}{r} 6 \\ +\ 9 \\ \hline 15 \end{array}$ 31. $\begin{array}{r} 3 \\ +\ 8 \\ \hline 11 \end{array}$

## EXERCISES 1.3

1. $5 + 7 = 12$
3. $7 + 5 = 12$
5. $(2 + 4) + 6 = 6 + 6$
   $= 12$
7. $2 + (4 + 6) = 2 + 10$
   $= 12$
9. $3 + 0 = 3$
11. $0 + 8 = 8$
13. $2 + 7 + 9 = 9 + 9$
    $= 18$
15. $2 + 3 + 4 + 9 = 5 + 4 + 9$
    $= 9 + 9$
    $= 18$
17. $5 + 8 = 8 + 5$ : commutative
19. $(4 + 5) + 8 = 4 + (5 + 8)$ : associative
21. $3 + (7 + 5) = (3 + 7) + 5$ : associative
23. $3 + (4 + 0) = 3 + 4$ : additive identity

## EXERCISES 1.4

1. $\begin{array}{r} 24 \\ +\ 3 \\ \hline 27 \end{array}$ 3. $\begin{array}{r} 23 \\ +\ 56 \\ \hline 79 \end{array}$ 5. $\begin{array}{r} 332 \\ +\ 54 \\ \hline 386 \end{array}$ 7. $\begin{array}{r} 307 \\ +\ 232 \\ \hline 539 \end{array}$ 9. $\begin{array}{r} 2792 \\ +\ 205 \\ \hline 2997 \end{array}$

11. $$\begin{array}{r} 2345 \\ +\ 6053 \\ \hline 8398 \end{array}$$

13. $$\begin{array}{r} 2531 \\ +\ 5354 \\ \hline 7885 \end{array}$$

15. $$\begin{array}{r} 21{,}314 \\ +\ 43{,}042 \\ \hline 64{,}356 \end{array}$$

17. $$\begin{array}{r} 13 \\ 21 \\ +\ 35 \\ \hline 69 \end{array}$$

19. $$\begin{array}{r} 3462 \\ 213 \\ +\ 24 \\ \hline 3699 \end{array}$$

21. $35 + 432 = 467$

23. $4 + 12 + 340 + 1213 = 1569$

25. The total of 23 and 31 = 23 + 31 = 54

27. The sum of 562 and 231 = 562 + 231 = 793

29. 34 more than 125 = 125 + 34 = 159

31. The sum of 23, 122, and 451
= 23 + 122 + 451 = 596

33. Total score = score on first nine + score on second nine
= 42 + 46
= 88

35. Score on second test = score on first test + 23
= 73 + 23
= 96

EXERCISES 1.5

1. $$\begin{array}{r} 47 \\ +\ 9 \\ \hline 56 \end{array}$$

3. $$\begin{array}{r} 23 \\ +\ 48 \\ \hline 71 \end{array}$$

5. $$\begin{array}{r} 31 \\ 27 \\ +\ 35 \\ \hline 93 \end{array}$$

7. $$\begin{array}{r} 213 \\ +\ 78 \\ \hline 291 \end{array}$$

9. $$\begin{array}{r} 703 \\ +\ 287 \\ \hline 990 \end{array}$$

11. $$\begin{array}{r} 589 \\ 306 \\ +\ 42 \\ \hline 937 \end{array}$$

13. $$\begin{array}{r} 590 \\ 345 \\ +\ 758 \\ \hline 1693 \end{array}$$

15. $$\begin{array}{r} 2578 \\ +\ 3455 \\ \hline 6033 \end{array}$$

17. $$\begin{array}{r} 3490 \\ 548 \\ +\ 25 \\ \hline 4063 \end{array}$$

19. $$\begin{array}{r} 2289 \\ 38 \\ 578 \\ +\ 3489 \\ \hline 6394 \end{array}$$

21. $$\begin{array}{r} 23{,}458 \\ +\ 32{,}623 \\ \hline 56{,}081 \end{array}$$

23. $$\begin{array}{r} 26{,}735 \\ 259 \\ 3{,}056 \\ +\ 35{,}489 \\ \hline 65{,}539 \end{array}$$

25. The sum of 79 and 735 = 79 + 735 = 814

27. The total of 38, 354, and 8 = 38 + 354 + 8 = 400

29. 23 + 2845 + 5 + 589 = 3462

31. The total of 2195, 348, 640, 59, and 23,785:

$$\begin{array}{r} 2{,}195 \\ 348 \\ 640 \\ 59 \\ +\ 23{,}785 \\ \hline 27{,}027 \end{array}$$

33. In 5, 12, 19, 26 each term, except first, is 7 plus the previous term:
5, 12, 19, 26, 26 + 7, 26 + 7 + 7, 26 + 7 + 7 + 7, 26 + 7 + 7 + 7 + 7

5, 12, 19, 26, $\underline{33}$, $\underline{40}$, $\underline{47}$, $\underline{54}$

35. In 7, 13, 19, 25 each term, except first, is 6 plus the previous term:
7, 13, 19, 25, 25 + 6, 25 + 6 + 6, 25 + 6 + 6 + 6, 25 + 6 + 6 + 6 + 6

7, 13, 19, 25, $\underline{31}$, $\underline{37}$, $\underline{43}$, $\underline{49}$

37. 90 + 55 + 56 + 110 + 18 + 11 + 157 = 497 North America
127 + 101 + 44 + 37 + 5 + 7 + 59 = 380 Latin America
186 + 23 + 19 + 15 + 2 + 2 + 25 = 272 Near East

## EXERCISES 1.6

1. 38, to nearest ten, 40

3. 253, to nearest ten, 250

5. 696, to nearest ten, 700

7. 3482, to nearest hundred, 3500

9. 5962, to nearest hundred, 6000

11. 4927, to nearest thousand, 5000

13. 23,429, to nearest thousand, 23,000

15. 787,000 to nearest ten thousand, 790,000

17. 21,800,000 to nearest million, 22,000,000

19. sum: $\begin{array}{r} 58 \\ 27 \\ +\ 33 \\ \hline 118 \end{array}$ estimate: $\begin{array}{r} 60 \\ 30 \\ +\ 30 \\ \hline 120 \end{array}$

21. sum: $\begin{array}{r} 87 \\ 53 \\ 41 \\ 93 \\ +\ 62 \\ \hline 336 \end{array}$ estimate: $\begin{array}{r} 90 \\ 50 \\ 40 \\ 90 \\ +\ 60 \\ \hline 330 \end{array}$

23. sum: $\begin{array}{r} 379 \\ 1215 \\ +\ 528 \\ \hline 2122 \end{array}$ estimate: $\begin{array}{r} 400 \\ 1200 \\ +\ 500 \\ \hline 2100 \end{array}$

25. sum: $\begin{array}{r} 1378 \\ 519 \\ 792 \\ +\ 2041 \\ \hline 4730 \end{array}$ estimate: $\begin{array}{r} 1400 \\ 500 \\ 800 \\ +\ 2000 \\ \hline 4700 \end{array}$

27. sum: $\begin{array}{r} 2238 \\ 3925 \\ +\ 5217 \\ \hline 11,380 \end{array}$ estimate: $\begin{array}{r} 2000 \\ 4000 \\ +\ 5000 \\ \hline 11,000 \end{array}$

29. sum: $\begin{array}{r} 9137 \\ 2315 \\ 7643 \\ +\ 3092 \\ \hline 22,187 \end{array}$ estimate: $\begin{array}{r} 9000 \\ 2000 \\ 8000 \\ +\ 3000 \\ \hline 22,000 \end{array}$

31. $4 < 8$

33. $500 > 400$

35. $100 < 1000$

37. 225,943,000,000 rounds to 226 billion
112,972,000,000 rounds to 113 billion

39. 118,596 rounds to 119,000
56,848 rounds to 57,000

## EXERCISES 1.7

1. In $9 - 6 = 3$, 9 is the minuend, 6 is the subtrahend, and 3 is the difference. $9 = 3 + 6$ is the related addition statement.

3. $\begin{array}{r} 86 \\ -\ 23 \\ \hline 63 \end{array}$ check: $\begin{array}{r} 63 \\ +\ 23 \\ \hline 86 \end{array}$

5. $97 - 45 = 52$
check: $45 + 52 = 97$

7. $98 - 57 = 41$
check: $57 + 41 = 98$

9.
$$\begin{array}{r} 347 \\ -\ 201 \\ \hline 146 \end{array}$$
check:
$$\begin{array}{r} 146 \\ +\ 201 \\ \hline 347 \end{array}$$

11.
$$\begin{array}{r} 689 \\ -\ 245 \\ \hline 444 \end{array}$$
check:
$$\begin{array}{r} 444 \\ +\ 245 \\ \hline 689 \end{array}$$

13.
$$\begin{array}{r} 3446 \\ -\ 2326 \\ \hline 1120 \end{array}$$
check:
$$\begin{array}{r} 1120 \\ +\ 2326 \\ \hline 3446 \end{array}$$

15.
$$\begin{array}{r} 8540 \\ -\ 2320 \\ \hline 6220 \end{array}$$
check:
$$\begin{array}{r} 6220 \\ +\ 2320 \\ \hline 8540 \end{array}$$

17.
$$\begin{array}{r} 23{,}689 \\ -\ 2{,}523 \\ \hline 21{,}166 \end{array}$$
check:
$$\begin{array}{r} 21{,}166 \\ +\ 2{,}523 \\ \hline 23{,}689 \end{array}$$

19.
$$\begin{array}{r} 47{,}235 \\ -\ 23{,}025 \\ \hline 24{,}210 \end{array}$$
check:
$$\begin{array}{r} 24{,}210 \\ +\ 23{,}025 \\ \hline 47{,}235 \end{array}$$

21. The difference of 97 and 43 = $97 - 43$
= 54

23. 25 less than 76 = $76 - 25$
= 51

25. 298 decreased by 47 = $298 - 47$
= 251

27. Tony's score = Danielle's score less 23
= $87 - 23$
= 64

29. Difference between large number and small number = 134
large number − small number = 134
655 − small number = 134
655 = small number + 134
655 − 134 = small number
521 = small number

EXERCISES 1.8

1.
$$\begin{array}{r} 64 \\ -\ 27 \\ \hline 37 \end{array}$$
check:
$$\begin{array}{r} 37 \\ +\ 27 \\ \hline 64 \end{array}$$

3.
$$\begin{array}{r} 50 \\ -\ 36 \\ \hline 14 \end{array}$$
check:
$$\begin{array}{r} 14 \\ +\ 36 \\ \hline 50 \end{array}$$

5.
$$\begin{array}{r} 372 \\ -\ 58 \\ \hline 314 \end{array}$$
check:
$$\begin{array}{r} 314 \\ +\ 58 \\ \hline 372 \end{array}$$

7.
$$\begin{array}{r} 534 \\ -\ 263 \\ \hline 271 \end{array}$$
check:
$$\begin{array}{r} 271 \\ +\ 263 \\ \hline 534 \end{array}$$

9. $\begin{array}{r} 627 \\ -\ 358 \\ \hline 269 \end{array}$ check: $\begin{array}{r} 269 \\ +\ 358 \\ \hline 627 \end{array}$

11. $\begin{array}{r} 280 \\ -\ 185 \\ \hline 95 \end{array}$ check: $\begin{array}{r} 95 \\ +\ 185 \\ \hline 280 \end{array}$

13. $\begin{array}{r} 603 \\ -\ 259 \\ \hline 344 \end{array}$ check: $\begin{array}{r} 344 \\ +\ 259 \\ \hline 603 \end{array}$

15. $\begin{array}{r} 2358 \\ -\ 562 \\ \hline 1796 \end{array}$ check: $\begin{array}{r} 1796 \\ +\ 562 \\ \hline 2358 \end{array}$

17. $\begin{array}{r} 3537 \\ -\ 2675 \\ \hline 862 \end{array}$ check: $\begin{array}{r} 862 \\ +\ 2675 \\ \hline 3537 \end{array}$

19. $\begin{array}{r} 6423 \\ -\ 3678 \\ \hline 2745 \end{array}$ check: $\begin{array}{r} 2745 \\ +\ 3678 \\ \hline 6423 \end{array}$

21. $\begin{array}{r} 6034 \\ -\ 2569 \\ \hline 3465 \end{array}$ check: $\begin{array}{r} 3465 \\ +\ 2569 \\ \hline 6034 \end{array}$

23. $\begin{array}{r} 4000 \\ -\ 2345 \\ \hline 1655 \end{array}$ check: $\begin{array}{r} 1655 \\ +\ 2345 \\ \hline 4000 \end{array}$

25. $\begin{array}{r} 33,486 \\ -\ 14,047 \\ \hline 19,439 \end{array}$ check: $\begin{array}{r} 19,439 \\ +\ 14,047 \\ \hline 33,486 \end{array}$

27. $\begin{array}{r} 29,400 \\ -\ 17,900 \\ \hline 11,500 \end{array}$ check: $\begin{array}{r} 11,500 \\ +\ 17,900 \\ \hline 29,400 \end{array}$

29. $\begin{array}{r} 59,000 \\ -\ 23,458 \\ \hline 35,542 \end{array}$ check: $\begin{array}{r} 35,542 \\ +\ 23,458 \\ \hline 59,000 \end{array}$

31.

| Beginning Balance | $351 |
|---|---|
| Check #1 | 29 |
| Balance | 322 |
| Check #2 | 139 |
| Balance | 183 |
| Check #3 | 75 |
| Ending Balance | 108 |

33. $\begin{array}{r} 5,846 \\ -\ 1,938 \\ \hline 3,908 \end{array}$ estimate: $\begin{array}{r} 6,000 \\ -\ 2,000 \\ \hline 4,000 \end{array}$ 3,908 is correct

35. $\begin{array}{r} 29,857 \\ -\ 2,098 \\ \hline 26,759 \end{array}$ estimate: $\begin{array}{r} 30,000 \\ -\ 2,000 \\ \hline 28,000 \end{array}$ 26,759 is incorrect (27,759 is correct)

37. 53, 47, 41, 35, <u>29</u>, <u>23</u>, <u>17</u>, <u>11</u> : pattern = subtract 6 from previous term.

39.

| | | | |
|---|---|---|---|
| | | | ↗ 8 + 5 + 2 = 15 |
| 4 | 9 | 2 | → 4 + 9 + 2 = 15 |
| 3 | 5 | 7 | → 3 + 5 + 7 = 15 |
| 8 | 1 | 6 | → 8 + 1 + 6 = 15 |
| | | | ↘ 4 + 5 + 6 = 15 |
| ↓ 4 3 + 8 | ↓ 9 5 + 1 | ↓ 2 7 + 6 | |
| 15 | 15 | 15 | |

sum of rows, columns, and diagonals is 15.

41. Label the missing numbers as shown, adding the diagonal from bottom left to top right.
$8 + 5 + 2 = 15$
The first row is $a + 7 + 2 = 15$ from which $a = 6$

| | | |
|---|---|---|
| a | 7 | 2 |
| b | 5 | 9 |
| 8 | d | e |

The first column is then $6 + b + 8 = 15$ from which $b = 1$.
The second column is $7 + 5 + d = 15$ from which $d = 3$.
The third row is $8 + 3 + e = 15$ from which $e = 4$. The square is now

| | | |
|---|---|---|
| 6 | 7 | 2 |
| 1 | 5 | 9 |
| 8 | 3 | 4 |

Checking shows all rows, columns, and diagonals add to 15.

43. Label the missing numbers as shown. Add along diagonal from top left to bottom right,
$16 + 10 + 7 + 1 = 34$.

| | | | |
|---|---|---|---|
| 16 | 3 | a | 13 |
| b | 10 | 11 | c |
| 9 | 6 | 7 | d |
| 4 | y | z | 1 |

From first row, $16 + 3 + a + 13 = 34$, $a = 2$.
From third row, $9 + 6 + 7 + d = 34$, $d = 12$.
From second column, $3 + 10 + 6 + y = 34$, $y = 15$.
From third column, $2 + 11 + 7 + z = 34$, $z = 14$.
From fourth column, $13 + c + 12 + 1 = 34$, $c = 8$.
From second row, $b + 10 + 11 + 8 = 34$, $b = 5$.

The final square is:

| | | | |
|---|---|---|---|
| 16 | 3 | 2 | 13 |
| 5 | 10 | 11 | 8 |
| 9 | 6 | 7 | 12 |
| 4 | 15 | 14 | 1 |

Checking shows all rows, columns, and diagonals add to 34.

45. 10,400 thousand
 − 6,900 thousand
 3,500 thousand metric tons

49. 145 thousand
 − 76 thousand
 69 thousand metric tons

EXERCISES 1.9

1. Distance driven = distance first day + distance second day
 = 385 + 273
 = 658 miles

3. Three-game total = score first game + score second game + score third game
 = 189 + 212 + 208
 = 609

5. Total number of tickets = \$9 seats + \$7 seats + \$5 seats
 = 245 + 350 + 475
 = 1070 tickets

7. Total number of passengers = sum of passengers on five flights
 = 137 + 179 + 154 + 201 + 168
 = 839 passengers

9. Ending balance = beginning balance − payment
 = \$543 − \$175
 = \$368

11. Margin = number of yes votes − number of no votes
 = 3457 − 3189
 = 268 votes

13. Difference in height = Sear's height − Empire height
 = 1454 − 1250
 = 204 ft

15.

| Regular Pay<br>+ Overtime Pay | 278<br>+ 53 | Taxes<br>+ Social Security | 49<br>+ 18 |
|---|---|---|---|
| Gross Pay | 331 | Deductions | 67 |

| Gross Pay<br>– Deductions | 331<br>– 67 | | |
|---|---|---|---|
| Net Pay | 264 | Take Home Pay | = $264 |

17. Total calories = Breakfast calories + lunch calories + dinner calories
= 270 + 450 + 820
= 1540

Difference between total calories and diet calories = amount over
1540 – 1500 = amount over
amount over = 40 calories

19. Sum of five tests = 540

95 + 84 + 82 + 89 + final exam score = 540
350 + final exam score = 540
final exam score = 540 – 350
final exam score = 190 points

21. Combined population = Sum of populations of three cities
= 1,600,000 + 880,500 + 872,500
= 3,353,000 people

# CHAPTER 2
# MULTIPLICATION OF WHOLE NUMBERS

## EXERCISES 2.1

1. $3 \times 7 = 7 + 7 + 7 = 14 + 7 = 21$

$$
\begin{aligned}
7 \times 3 &= 3 + 3 + 3 + 3 + 3 + 3 + 3 \\
&= 6 + 3 + 3 + 3 + 3 + 3 \\
&= 9 + 3 + 3 + 3 + 3 \\
&= 12 + 3 + 3 + 3 \\
&= 15 + 3 + 3 \\
&= 18 + 3 \\
&= 21
\end{aligned}
$$

3. In $6 \times 7 = 42$, 6 and 7 are called *factors* of 42. In $6 \times 7 = 42$, 42 is called the *product* of 6 and 7.

5. The factors of 30 are 1, 2, 3, 5, 6, 10, 15, 30.

7. The numbers 3, 6, 9, 12, 15, $\cdots$ are *multiples* of 3.

9. $\begin{array}{r} 5 \\ \times\ 3 \\ \hline 15 \end{array}$ 11. $\begin{array}{r} 8 \\ \times\ 1 \\ \hline 8 \end{array}$ 13. $\begin{array}{r} 6 \\ \times\ 0 \\ \hline 0 \end{array}$ 15. $\begin{array}{r} 2 \\ \times\ 9 \\ \hline 18 \end{array}$ 17. $\begin{array}{r} 5 \\ \times\ 6 \\ \hline 30 \end{array}$

19. $\begin{array}{r} 4 \\ \times\ 9 \\ \hline 36 \end{array}$ 21. $\begin{array}{r} 3 \\ \times\ 8 \\ \hline 24 \end{array}$ 23. $\begin{array}{r} 5 \\ \times\ 7 \\ \hline 35 \end{array}$ 25. $\begin{array}{r} 6 \\ \times\ 9 \\ \hline 54 \end{array}$ 27. $\begin{array}{r} 8 \\ \times\ 8 \\ \hline 64 \end{array}$

29. $\begin{array}{r} 9 \\ \times\ 8 \\ \hline 72 \end{array}$ 31. $\begin{array}{r} 5 \\ \times\ 8 \\ \hline 40 \end{array}$ 33. $\begin{array}{r} 4 \\ \times\ 6 \\ \hline 24 \end{array}$ 35. $\begin{array}{r} 3 \\ \times\ 9 \\ \hline 27 \end{array}$

37. $5 \cdot 9 = 45$

39. $(4)(8) = 32$

41. $(9)(6) = 54$

43. $(3)(2)(4) = (6)(4)$
$= 24$

## EXERCISES 2.2

1. $5 \times 8 = 8 \times 5$, commutative property of multiplication.

3. $5 \times 0 = 0$, multiplication property of zero.

5. $2 \times (3 \times 5) = (2 \times 3) \times 5$, associative property of multiplication.

7. $9 \times 3 = 3 \times 9$, commutative property of multiplication.

9. $1 \times 5 = 5$, multilicative identity.

11. $0 \times 9 = 0$, multiplication property of zero.

13. $5 \times (2 \times 3) = (5 \times 2) \times 3$, associative property of multiplication.

15. $3 \times (2 + 8) = (3 \times 2) + (3 \times 8)$, distributive property.

EXERCISES 2.3

1.
23
× 2
46

3.
48
× 4
192

5.
508
× 6
3048

7.
523
× 8
4184

9.
2035
× 9
18,315

11.
5478
× 7
38,346

13.
26,555
× 7
185,885

15.
20,108
× 7
140,756

17. 245 × 8 = 1960

19. 3249 × 5 = 16,245

21. Product of 304 and 7 = 304 × 7 = 2128

23. Product of 8 and 5679 = 8 × 5679 = 45,432

25. 18 × 6 = 108 packs

EXERCISES 2.4

1.
47
× 38
376
141
1786

3.
98
× 57
686
490
5586

5.
235
× 49
2115
940
11,515

7.
2364
× 67
16 548
141 84
158,388

9.
315
× 243
945
1260
630
76,545

11.
345
× 267
2 415
20 70
69 0
92,115

13.
547
× 203
1 641
109 4
111,041

15.
2 458
× 135
12 290
73 74
245 8
331,830

17.
1 208
× 305
6 040
362 4
368,440

19.
2 534
× 3 106
15 204
253 4
7 602
7,870,604

21. The product of 203 and 57.
203
× 57
1 421
10 15
11,571

23. The product of 135 and 507:
135
× 507
945
67 5
68,445

25. 84 × 12 = $1008

27. 34 × 27 = $918

EXERCISES 2.5

1.
53
× 10
530

3.
89
× 100
8,900

5.
567
× 1000
567,000

7.
236
× 10,000
2,360,000

9.
43
× 70
3010

11.
562
× 400
224,800

13.
3 45
× 230
10 350
69 0
79,350

15.
15 7
× 3200
31 400
471
502,400

17. $\begin{array}{r} 3\ 67 \\ \times\ \ 20 \\ \hline 7{,}340 \end{array}$

19. $\begin{array}{r} 2\ 49 \\ \times\ \ 300 \\ \hline 74{,}700 \end{array}$

21. $\begin{array}{r} 238 \\ \times\ \ 4000 \\ \hline 952{,}000 \end{array}$

23. $\begin{array}{r} 408 \\ \times\ \ 5000 \\ \hline 2{,}040{,}000 \end{array}$

25. $\begin{array}{r} 36 \\ \times\ 23 \\ \hline 828 \end{array}$ estimate: $\begin{array}{r} 40 \\ \times\ 20 \\ \hline 800 \end{array}$

27. $\begin{array}{r} 93 \\ \times\ 48 \\ \hline 4464 \end{array}$ estimate: $\begin{array}{r} 90 \\ \times\ 50 \\ \hline 4500 \end{array}$

29. $\begin{array}{r} 212 \\ \times\ 278 \\ \hline 58{,}936 \end{array}$ estimate: $\begin{array}{r} 200 \\ \times\ \ 300 \\ \hline 60{,}000 \end{array}$

31. $\begin{array}{r} 391 \\ \times\ 531 \\ \hline 207{,}621 \end{array}$ estimate: $\begin{array}{r} 400 \\ \times\ \ 500 \\ \hline 200{,}000 \end{array}$

33. $250{,}000{,}000 \times 2162 = 540{,}500{,}000{,}000$ cubic meters

35. $1{,}251{,}000 \times 25 = 31{,}275{,}000$ hectares

## EXERCISES 2.6

1. $4 \times 5 + 7 = 20 + 7$
   $= 27$

3. $3 + 6 \times 4 = 3 + 24$
   $= 27$

5. $8 \times 5 - 20 = 40 - 20$
   $= 20$

7. $48 - 8 \times 5 = 48 - 40$
   $= 8$

9. $9 + 6 \times 8 = 9 + 48$
   $= 57$

11. $20 \times 6 - 5 = 120 - 5$
   $= 115$

13. $20 \times (6 - 5) = 20 \times (1)$
   $= 20$

15. $9 \times (4 + 3) = 9 \times 7$
   $= 63$

17. $4 \times 5 + 7 \times 3 = 20 + 21$
   $= 41$

19. $9 \times 8 - 12 \times 6 = 72 - 72$
   $= 0$

21. $5 \times (3 + 4) = 5 \times (7)$
   $= 35$

23. $5 \times 3 + 5 \times 4 = 15 + 20$
   $= 35$

## EXERCISES 2.7

1. $\begin{array}{r} 850 \\ \times\ 8 \\ \hline 6800 \end{array}$ sq ft

3. $\begin{array}{r} 125 \\ \times\ 12 \\ \hline 250 \\ 125\ \ \\ \hline \$1500 \end{array}$

5. $\begin{array}{r} 24 \\ \times\ 18 \\ \hline 192 \\ 24\ \ \\ \hline 432 \end{array}$ pictures

7. $\begin{array}{r} 40 \\ \times\ \ 60 \\ \hline 2400 \end{array}$ labels

9. $\begin{array}{r} 879 \\ \times\ \ 28 \\ \hline 7\ 032 \\ 17\ 58\ \ \\ \hline \$24{,}612 \end{array}$

11. $\begin{array}{r} 1\ 088 \\ \times\ \ \ 23 \\ \hline 3\ 264 \\ 21\ 76\ \ \\ \hline 25{,}024 \end{array}$ feet

13. area = 

$$\begin{array}{r} 35 \\ \times\ 45 \\ \hline 175 \\ 140\phantom{0} \\ \hline 1575 \end{array}$$ sq ft

15. area = 

$$\begin{array}{r} 24 \\ \times\ 28 \\ \hline 192 \\ 48\phantom{0} \\ \hline 672 \end{array}$$ sq ft

17. Cost = cost per sq yd × area. Area = 4 × 5 = 20
= 13 × 20
= \$260

19. Cost = down payment + monthly payment × number of months
= 125 + 25 × 12
= 125 + 300
= \$425

21. Total area = sum of individual areas
= 225 × 250 + 300 × 275
= 56,250 + 82,500
= 138,750 sq ft

23. Yearly earnings = \$412 per week for 22 weeks + (412 + 70) per week for 30 weeks
= 412 × 22 + (412 + 70) × 30
= 9064 + (482) × 30
= 9064 + 14,460
= \$23,524

25. Number of ways = 16 × 15 × 14 × 13 = 43,680

27. Minutes per year = (60 min per hr) × (24 hr per day) × (365 days per year)
= 525,600 min in one year

EXERCISES 2.8

1. $3^2 = 3 \times 3 = 9$

3. $2^4 = 2 \times 2 \times 2 \times 2 = 16$

5. $8^3 = 8 \times 8 \times 8 = 512$

7. $1^5 = 1 \times 1 \times 1 \times 1 \times 1 = 1$

9. $5^1 = 5$

11. $9^0 = 1$

13. $10^3 = 10 \times 10 \times 10 = 1000$

15. $10^6 = 10 \times 10 \times 10 \times 10 \times 10 \times 10 = 1{,}000{,}000$

17. $2 \times 4^3 = 2 \times 4 \times 4 \times 4 = 128$

19. $2 \times 3^2 = 2 \times 3 \times 3 = 18$

21. $5 + 2^2 = 5 + 2 \times 2 = 5 + 4 = 9$

23. $(3 \times 2)^4 = 6^4 = 6 \times 6 \times 6 \times 6 = 1296$

25. $2 \times 6^2 = 2 \times 6 \times 6 = 72$

27. $14 - 3^2 = 14 - 3 \times 3 = 14 - 9 = 5$

29. $(3 + 2)^3 - 20 = 5^3 - 20 = 5 \times 5 \times 5 - 20 = 125 - 20 = 105$

31. $(7 - 4)^4 - 30 = (3)^4 - 30 = 3 \times 3 \times 3 \times 3 - 30 = 81 - 30 = 51$

33. 6, 8, 10 : $6^2 = 6 \times 6 = 36$ $\quad 36 + 64 = 100$
$8^2 = 8 \times 8 = 64$
$10^2 = 10 \times 10 = 100$
6, 8, 10 are Pythagorean triples

35. 5, 12, 13: $5^2 = 5 \times 5 = 25$ $\quad 25 + 144 = 169$

$12^2 = 12 \times 12 = 144$

$13^2 = 13 \times 13 = 169$

5, 12, 13 are Pythagorean triples.

37. 8, 16, 18: $8^2 = 8 \times 8 = 64$ $\quad 64 + 256 = 320 \neq 324$

$16^2 = 16 \times 16 = 256$

$18^2 = 18 \times 18 = 324$

8, 16, 18 are not Pythagorean triples.

# CHAPTER 3
# DIVISION OF WHOLE NUMBERS

## EXERCISES 3.1

1. In $48 \div 8 = 6$, 8 is the *divisor*, 48 is the *dividend*, and 6 is the *quotient*

3. $\begin{array}{r} 36 \\ -\ 9 \\ \hline 27 \end{array}$ $\quad \begin{array}{r} 27 \\ -\ 9 \\ \hline 18 \end{array}$ $\quad \begin{array}{r} 18 \\ -\ 9 \\ \hline 9 \end{array}$ $\quad \begin{array}{r} 9 \\ -\ 9 \\ \hline 0 \end{array}$ $\quad 36 \div 9 = 4$

5. The division $30 \div 6 = 5$ is $30 = 6 \times 5$ as a multiplication

7. $\begin{array}{r} 8\ r3 \\ 4\overline{)35} \end{array}$ $\quad 4 \times 8 + 3 = 32 + 3 = 35$

9. $\begin{array}{r} 5 \\ 7\overline{)35} \end{array}$ $\quad 7 \times 5 = 35$

11. $\begin{array}{r} 8 \\ 5\overline{)40} \end{array}$ $\quad 5 \times 8 = 40$

13. $\begin{array}{r} 7 \\ 3\overline{)21} \end{array}$ $\quad 3 \times 7 = 21$

15. $\begin{array}{r} 9 \\ 7\overline{)63} \end{array}$ $\quad 7 \times 9 = 63$

17. $\begin{array}{r} 7 \\ 8\overline{)56} \end{array}$ $\quad 8 \times 7 = 56$

19. $\begin{array}{r} 12 \\ 7\overline{)84} \end{array}$ $\quad 7 \times 12 = 84$

21. $\begin{array}{r} 5\ r1 \\ 3\overline{)16} \\ 15 \\ \hline 1 \end{array}$ $\quad 3 \times 5 + 1 = 15 + 1 = 16$

23. $\begin{array}{r} 3\ r3 \\ 8\overline{)27} \\ 24 \\ \hline 3 \end{array}$ $\quad 8 \times 3 + 3 = 24 + 3 = 27$

25. $\begin{array}{r} 5\ r4 \\ 7\overline{)39} \\ 35 \\ \hline 4 \end{array}$ $\quad 7 \times 5 + 4 = 35 + 4 = 39$

27. $\begin{array}{r} 8\ r3 \\ 5\overline{)43} \\ 40 \\ \hline 3 \end{array}$ $\quad 5 \times 8 + 3 = 40 + 3 = 43$

29. $\begin{array}{r} 7\ r2 \\ 9\overline{)65} \\ 63 \\ \hline 2 \end{array}$ $\quad 9 \times 7 + 2 = 63 + 2 = 65$

31. $\begin{array}{r} 7\ r1 \\ 8\overline{)57} \\ 56 \\ \hline 1 \end{array}$ $\quad 8 \times 7 + 1 = 56 + 1 = 57$

33. $\begin{array}{r} 4 \\ 15\overline{)60} \\ 60 \\ \hline 0 \end{array}$ classrooms

35. $\begin{array}{r} 7 \\ 8\overline{)56} \\ 56 \\ \hline 0 \end{array}$ boxes

## EXERCISES 3.2

1. $\begin{array}{r} 1 \\ 5\overline{)5} \end{array}$

3. $0 \div 5 = 0$

5. $\begin{array}{r} 1 \\ 5\overline{)5} \end{array}$

7. $\begin{array}{r} 6 \\ 1\overline{)6} \end{array}$

9. $9 \div 9 = 1$

11. $4 \div 0$, undefined

13. $\begin{array}{r} 1 \\ 8\overline{)8} \end{array}$

15. $8 \div 1 = 8$

17. $0 \div 6 = 0$

19. $10 \div 1 = 10$

## EXERCISES 3.3

1. $$\begin{array}{r} 16\ r3 \\ 5\overline{)83} \\ \underline{5}\phantom{3} \\ 33 \\ \underline{30} \\ 3 \end{array}$$

$5 \times 16 + 3 = 80 + 3 = 83$

3. $$\begin{array}{r} 13\ r2 \\ 7\overline{)93} \\ \underline{7}\phantom{3} \\ 23 \\ \underline{21} \\ 2 \end{array}$$

$7 \times 13 + 2 = 91 + 2 = 93$

5. $$\begin{array}{r} 54 \\ 3\overline{)162} \\ \underline{15}\phantom{2} \\ 12 \\ \underline{12} \\ 0 \end{array}$$

$3 \times 54 = 162$

7. $$\begin{array}{r} 36\ r5 \\ 8\overline{)293} \\ \underline{24}\phantom{3} \\ 53 \\ \underline{48} \\ 5 \end{array}$$

$8 \times 36 + 5 = 288 + 5 = 293$

9. $$\begin{array}{r} 147\ r3 \\ 5\overline{)738} \\ \underline{5}\phantom{38} \\ 23\phantom{8} \\ \underline{20}\phantom{8} \\ 38 \\ \underline{35} \\ 3 \end{array}$$

$5 \times 147 + 3 = 735 + 3 = 738$

11. $$\begin{array}{r} 121\ r7 \\ 8\overline{)975} \\ \underline{8}\phantom{75} \\ 17\phantom{5} \\ \underline{16}\phantom{5} \\ 15 \\ \underline{8} \\ 7 \end{array}$$

$8 \times 121 + 7 = 968 + 7 = 975$

13. $$\begin{array}{r} 108\ r1 \\ 6\overline{)649} \\ \underline{6}\phantom{49} \\ 49 \\ \underline{48} \\ 1 \end{array}$$

$6 \times 108 + 1 = 648 + 1 = 649$

15. $$\begin{array}{r} 392 \\ 8\overline{)3136} \\ \underline{24}\phantom{36} \\ 73\phantom{6} \\ \underline{72}\phantom{6} \\ 16 \\ \underline{16} \\ 0 \end{array}$$

$8 \times 392 = 3136$

17. $$\begin{array}{r} 679\ \ r6 \\ 8\overline{)5438} \\ \underline{48}\phantom{38} \\ 63\phantom{8} \\ \underline{56}\phantom{8} \\ 78 \\ \underline{72} \\ 6 \end{array}$$

$8 \times 679 + 6 = 5432 + 6 = 5438$

19. $$\begin{array}{r} 1451 \\ 4\overline{)5804} \\ \underline{4}\phantom{804} \\ 18\phantom{04} \\ \underline{16}\phantom{04} \\ 20\phantom{4} \\ \underline{20}\phantom{4} \\ 4 \\ \underline{4} \\ 0 \end{array}$$

$4 \times 1451 = 5804$

21. $5\overline{)8643}$ quotient 1728 r3

```
   1728  r3
5)8643
  5
  36
  35
   14
   10
    43
    40
     3
```

$5 \times 1728 + 3 = 8640 + 3$
$= 8643$

23.

```
   307  r4
7)2153
  21
   53
   49
    4
```

$7 \times 307 + 4 = 2149 + 4$
$= 2153$

25.

```
   2769  r1
8)22,153
  16
   61
   56
    55
    48
     73
     72
      1
```

$8 \times 2769 + 1 = 22,152 + 1$
$= 22,153$

27.

```
   11716  r1
7)82,013
  7
  12
   7
   50
   49
    11
     7
     43
     42
      1
```

$7 \times 11,716 + 1 = 82,012 + 1$
$= 82,013$

29.

```
   451  r2
6)2708
  24
   30
   30
    8
    6
    2
```

remainder is 2

31. quotient = 451 (from 29)

33.

```
   9
8)77
  72
   5
```

5 pictures will be left over

EXERCISES 3.4

1.

```
      5  r55
58)345
   290
    55
```

$58 \times 5 + 55 = 290 + 55$
$= 345$

3.

```
      6  r51
63)429
   378
    51
```

$63 \times 6 + 51 = 378 + 51$
$= 429$

5.

```
     18  r28
48)892
   48
   412
   384
    28
```

$48 \times 18 + 28 = 864 + 28$
$= 892$

7.

```
     23  r5
23)534
   46
    74
    69
     5
```

$23 \times 23 + 5 = 529 + 5$
$= 534$

9.

```
     52  r27
45)2367
   225
   117
    90
    27
```

$45 \times 52 + 27 = 2340 + 27$
$= 2367$

11.

```
    257 r10
34)8748
   68
   194
   170
    248
    238
     10
```

$34 \times 257 + 10 = 8738 + 10$
$= 8748$

13.

```
    189 r14
42)7952
   42
   375
   336
    392
    378
     14
```

$42 \times 189 + 14 = 7938 + 14$
$= 7952$

15.

```
    305  r7
28)8547
   84
    147
    140
      7
```

$28 \times 305 + 7 = 8540 + 7$
$= 8547$

17.

```
        5 r56
763)3871
    3815
      56
```

$763 \times 5 + 56 = 3815 + 56$
$= 3871$

19.

```
      23 r66
326)7564
    652
    1044
     978
      66
```

$326 \times 23 + 66 = 7498 + 66$
$= 7564$

21.

```
      20 r130
432)8770
    8640
     130
```

$432 \times 20 + 130 = 8640 + 130$
$= 8770$

23.

```
       72 r63
454)32751
    3178
     971
     908
      63
```

$454 \times 72 + 63 = 32,688 + 63$
$= 32,751$

25.

```
      205   r70
103)21185
    206
      585
      515
       70
```

$103 \times 205 + 70 = 21,115 + 70$
$= 21,185$

27.

```
       534 r44
234)125000
    1170
      800
      702
       980
       936
        44
```

$234 \times 534 + 44 = 124956 + 44$
$= 125,000$

EXERCISES 3.5

1.
```
   21  r1
4)85
```

3.
```
   29  r1
3)88
```

5.
```
   212  r1
4)848
```

7.
```
   48 r6
7)342
```

9.
```
   104 r1
6)625
```

11.
```
   158 r2
4)634
```

13.
```
   472  r4
5)2364
```

15.
```
   1087  r3
4)4351
```

17.
```
   1830  r1
4)7321
```

19.
```
   408  r5
6)2453
```

21.
```
   4473 r2
3)13,421
```

23.
```
   4589 r5
7)32,128
```

EXERCISES 3.6

1. $8 \div 4 + 2 = 2 + 2$
$= 4$

3. $24 - 6 \div 3 = 24 - 2$
$= 22$

5. $(24 - 6) \div 3 = 18 \div 3$
$= 6$

7. $12 + 3 \div 3 = 12 + 1$
$= 13$

9. $18 \div 6 \times 3 = 3 \times 3$
$= 9$

11. $30 \div 6 - 12 \div 3 = 5 - 4$
$= 1$

13. $4^2 \div 2 = 16 \div 2$
$= 8$

15. $5^2 \times 3 = 25 \times 3$
$= 75$

17. $3 \times 3^3 = 3 \times 27$
$= 81$

19. $(3^3 + 3) \div 10 = (27 + 3) \div 10$
$= 30 \div 10$
$= 3$

21. $15 \div (5 - 3 + 1) = 15 \div (2 + 1)$
$= 15 \div 3$
$= 5$

23. $27 \div (2^2 + 5) = 27 \div (4 + 5)$
$= 27 \div 9$
$= 3$

EXERCISES 3.7

1. receipts = number of tickets × price per ticket
552 = number of tickets × 4

$$\text{number of tickets} = \frac{552}{4}$$
= 138 tickets

3. $$\text{average} = \frac{\text{total number of points}}{\text{total number of games}} = \frac{476}{17}$$
average = 28 points per game

5. $$\text{average enrollment} = \frac{\text{total number of students}}{\text{total number of sections}} = \frac{522}{18}$$
average enrollment = 29 students per section

7. $$\text{owner cost} = \frac{\text{total cost}}{\text{number of owners}} = \frac{2030}{14}$$
owner cost = $145 per owner

9. $$\text{average number of calls} = \frac{\text{total number of calls}}{\text{total number of phones}} = \frac{1702}{37}$$
average number of calls = 46 calls per phone

11. number of lines printed = number of lines per minute × number of minutes
10880 = 340 × time

$$\text{time} = \frac{10880}{340} = 32 \text{ minutes}$$

13. $$\text{bonus} = \frac{\text{total amount distributed}}{\text{number of employees}} = \frac{16{,}488}{36} = 458$$
Each employee receives a bonus of $458

15. $$\text{payment} = \frac{\text{cost} - \text{down payment}}{\text{number of months}} = \frac{9852 - 1500}{36}$$
$$= \frac{8352}{36}$$
payment = $232 per month

17. amount covered = amount per gallon × number of gallons

$$135 \times 20 = 450 \times \text{number of gallons}$$

$$2700 = 450 \times \text{number of gallons}$$

$$\text{number of gallons} = \frac{2700}{450}$$

number of gallons = 6

19. $9\overline{)97531}$ = 10836 r7  $9\overline{)35197}$ = 3910 r7  $9\overline{)59137}$ = 6570 r7

The remainder is 7 in all three cases.

21. $\text{population per square mile} = \frac{\text{population}}{\text{area}} = \frac{3{,}078{,}000}{114{,}000}$

population per square mile = 27 people per square mile

EXERCISES 3.8

1. $\text{average} = \frac{8 + 12 + 13}{3} = \frac{33}{3} = 11$

3. $\text{average} = \frac{5 + 8 + 11 + 12}{4} = \frac{36}{4} = 9$

5. $\text{average} = \frac{23 + 34 + 25 + 19 + 31 + 24}{6} = \frac{156}{6} = 26$

7. $\text{average temperature} = \frac{86 + 91 + 92 + 103 + 98}{5} = \frac{470}{5} = 94^\circ$

9. $\text{average rating} = \frac{43 + 29 + 51 + 36 + 33 + 42 + 32}{7} = \frac{266}{7}$ = 38 miles per gallon

11. $\text{average} = \frac{\text{sum of five tests}}{5}$

$$90 = \frac{83 + 93 + 88 + 91 + \text{last test score}}{5}$$

$$90 = \frac{355 + \text{last test score}}{5}$$

$$90 \times 5 = 355 + \text{last test score}$$

$$450 = 355 + \text{last test score}$$

$$450 - 355 = \text{last test score}$$

last test score = 95

13. $\text{Lewis' average} = \frac{87 + 82 + 93 + 89 + 84}{5} = \frac{435}{5} = 87$

$\text{Cheryl's average} = \frac{92 + 83 + 89 + 94 + 87}{5} = \frac{445}{5} = 89$

Cheryl's average was 2 points higher than Lewis' average.

15. average number of kWh for heating $= \dfrac{1200 + 1086 + 1103 + 975}{4} = 1091$ kWh

17. average number of kWh per appliance $= \dfrac{115 + 1086 + 386 + 154 + 99 + 117 + 45}{7} = 286$ kWh

19. average monthly bill $= \dfrac{53 + 51 + 43 + 37 + 32 + 29 + 34 + 41 + 58 + 55 + 49 + 58}{12} = \$45$

# CHAPTER 4
# FACTORS AND MULTIPLES

## EXERCISES 4.1

1. The factors of 4 are 1, 2, 4.

3. The factors of 10 are 1, 2, 5, 10.

5. The factors of 15 are 1, 3, 5, 15.

7. The factors of 24 are 1, 2, 3, 4, 6, 8, 12, 24.

9. The factors of 64 are 1, 2, 4, 8, 16, 32, 64.

11. The factors of 11 are 1, 11.

13. 13, 19, 23, 31, 59, 97 and 103 are all prime numbers

15. 31, 37, 41, 43, 47 are the prime numbers between 30 and 50.

17. 72, 158, 260, 378, 570, 4530, and 8300 are divisible by 2.

19. 45, 260, 570, 585, and 4530 are divisible by 3.

## EXERCISES 4.2

1. $18 = 2 \cdot 3 \cdot 3$

3. $30 = 2 \cdot 3 \cdot 5$

5. $51 = 3 \cdot 17$

7. $63 = 3 \cdot 3 \cdot 7$

9. $70 = 2 \cdot 5 \cdot 7$

11. $66 = 2 \cdot 3 \cdot 11$

13. $130 = 2 \cdot 5 \cdot 13$

15. $315 = 3 \cdot 3 \cdot 5 \cdot 7$

17. $225 = 3 \cdot 3 \cdot 5 \cdot 5$

19. $189 = 3 \cdot 3 \cdot 3 \cdot 7$

21. $336 = 2 \cdot 2 \cdot 2 \cdot 2 \cdot 3 \cdot 7$

23. $840 = 2 \cdot 2 \cdot 2 \cdot 3 \cdot 5 \cdot 7$

25. 24 has factors of 1, 2, 3, 4, 6, 8, 12, 24. Two factors with a sum of 10 are 6 and 4, $6 + 4 = 10$.

27. 30 has factors of 1, 2, 3, 5, 6, 10, 15, 30. Two factors with a difference of 1 are 6 and 5, $6 - 5 = 1$.

## EXERCISES 4.3

1. $4 = 2 \cdot 2$, $6 = 2 \cdot 3$; gcf $= 2$

3. $10 = 2 \cdot 5$, $15 = 3 \cdot 5$; gcf $= 5$

5. $21 = 3 \cdot 7$, $24 = 2 \cdot 2 \cdot 2 \cdot 3$; gcf $= 3$

7. $20 = 2 \cdot 2 \cdot 5$, $21 = 3 \cdot 7$; gcf $= 1$

9. $18 = 2 \cdot 3 \cdot 3$, $24 = 2 \cdot 2 \cdot 2 \cdot 3$; gcf $= 2 \cdot 3 = 6$

11. $18 = 2 \cdot 3 \cdot 3$, $54 = 2 \cdot 3 \cdot 3 \cdot 3$; gcf $= 2 \cdot 3 \cdot 3 = 18$

13. $36 = 2 \cdot 2 \cdot 3 \cdot 3$, $48 = 2 \cdot 2 \cdot 2 \cdot 2 \cdot 3$; gcf $= 2 \cdot 2 \cdot 3 = 12$

15. $84 = 2 \cdot 2 \cdot 3 \cdot 7$, $105 = 3 \cdot 5 \cdot 7$; gcf $= 3 \cdot 7 = 21$

17. $45 = 3 \cdot 3 \cdot 5$, $60 = 2 \cdot 2 \cdot 3 \cdot 5$, $75 = 3 \cdot 5 \cdot 5$; gcf $= 3 \cdot 5 = 15$

19. $12 = 2 \cdot 2 \cdot 3$, $36 = 2 \cdot 2 \cdot 3 \cdot 3$, $60 = 2 \cdot 2 \cdot 3 \cdot 5$; gcf $= 2 \cdot 2 \cdot 3 = 12$

21. $105 = 3 \cdot 5 \cdot 7$, $140 = 2 \cdot 2 \cdot 5 \cdot 7$, $175 = 5 \cdot 5 \cdot 7$; gcf $= 5 \cdot 7 = 35$

23. $25 = 5 \cdot 5$, $75 = 3 \cdot 5 \cdot 5$, $150 = 2 \cdot 3 \cdot 5 \cdot 5$; gcf $= 5 \cdot 5 = 25$

EXERCISES 4.4

1. $2 = 2$ $\quad$ $3 = 3$ $\quad$ LCM $= 2\cdot 3 = 6$

3. $4 = 2\cdot 2$ $\quad$ $6 = 2\cdot 3$ $\quad$ LCM $= 2\cdot 2\cdot 3 = 12$

5. $10 = 2\cdot 5$ $\quad$ $20 = 2\cdot 2\cdot 5$ $\quad$ LCM $= 2\cdot 2\cdot 5 = 20$

7. $9 = 3\cdot 3$ $\quad$ $12 = 2\cdot 2\cdot 3$ $\quad$ LCM $= 2\cdot 2\cdot 3\cdot 3 = 36$

9. $12 = 2\cdot 2\cdot 3$ $\quad$ $16 = 2\cdot 2\cdot 2\cdot 2$ $\quad$ LCM $= 2\cdot 2\cdot 2\cdot 2\cdot 3 = 48$

11. $12 = 2\cdot 2\cdot 3$ $\quad$ $15 = 3\cdot 5$ $\quad$ LCM $= 2\cdot 2\cdot 3\cdot 5 = 60$

13. $18 = 2\cdot 3\cdot 3$ $\quad$ $36 = 2\cdot 2\cdot 3\cdot 3$ $\quad$ LCM $= 2\cdot 2\cdot 3\cdot 3 = 36$

15. $25 = 5\cdot 5$ $\quad$ $40 = 2\cdot 2\cdot 2\cdot 5$ $\quad$ LCM $= 2\cdot 2\cdot 2\cdot 5\cdot 5 = 200$

17. $30 = 2\cdot 3\cdot 5$ $\quad$ $40 = 2\cdot 2\cdot 2\cdot 5$ $\quad$ LCM $= 2\cdot 2\cdot 2\cdot 3\cdot 5 = 120$

19. $8 = 2\cdot 2\cdot 2$ $\quad$ $15 = 3\cdot 5$ $\quad$ LCM $= 2\cdot 2\cdot 2\cdot 3\cdot 5 = 120$

21. $30 = 2\cdot 3\cdot 5$ $\quad$ $150 = 2\cdot 3\cdot 5\cdot 5$ $\quad$ LCM $= 2\cdot 3\cdot 5\cdot 5 = 150$

23. $8 = 2\cdot 2\cdot 2$ $\quad$ $48 = 2\cdot 2\cdot 2\cdot 2\cdot 3$ $\quad$ LCM $= 2\cdot 2\cdot 2\cdot 2\cdot 3 = 48$

25. $2 = 2$ $\quad$ $3 = 3$ $\quad$ $5 = 5$ $\quad$ LCM $= 2\cdot 3\cdot 5 = 30$

27. $3 = 3$ $\quad$ $5 = 5$ $\quad$ $6 = 2\cdot 3$ $\quad$ LCM $= 2\cdot 3\cdot 5 = 30$

29. $18 = 2\cdot 3\cdot 3$ $\quad$ $21 = 3\cdot 7$ $\quad$ $28 = 2\cdot 2\cdot 7$ $\quad$ LCM $= 2\cdot 2\cdot 3\cdot 3\cdot 7 = 252$

31. $20 = 2\cdot 2\cdot 5$ $\quad$ $30 = 2\cdot 3\cdot 5$ $\quad$ $45 = 3\cdot 3\cdot 5$ $\quad$ LCM $= 2\cdot 2\cdot 3\cdot 3\cdot 5 = 180$

## CHAPTER 5
## AN INTRODUCTION TO FRACTIONS

EXERCISES 5.1

1. $\frac{6}{11}$, numerator = 6; denominator = 11

3. $\frac{3}{11}$, numerator = 3; denominator = 11

5. $\frac{3}{4}$ shaded

7. $\frac{5}{6}$ shaded

9. $\frac{5}{5}$ or 1 shaded

11. $\frac{11}{12}$ shaded

13. $\frac{7}{12}$ shaded

15. $\frac{4}{5}$ shaded

17. $\frac{5}{8}$ shaded

19. Seven of twenty missed: $\frac{13}{20}$ correct; $\frac{7}{20}$ incorrect

21. 11 of 17 sold: $\frac{11}{17}$ sold; $\frac{6}{17}$ not sold.

23. $\frac{2}{5}$ may also be written as $2 \div 5$.

## EXERCISES 5.2

1. $\frac{3}{5}$, proper
3. $2\frac{3}{5}$, mixed number
5. $\frac{6}{6}$, improper
7. $\frac{11}{8}$, improper
9. $4\frac{5}{7}$, mixed number
11. $\frac{13}{17}$, proper
13. $\frac{31}{25}$, improper
15. $6\frac{2}{5}$, mixed number
17. $1\frac{3}{4}$ shaded
19. $3\frac{5}{8}$ shaded

## EXERCISES 5.3

1. $\frac{7}{2} = 3\frac{1}{2}$
3. $\frac{5}{4} = 1\frac{1}{4}$
5. $\frac{22}{5} = 4\frac{2}{5}$
7. $\frac{34}{5} = 6\frac{4}{5}$
9. $\frac{59}{5} = 11\frac{4}{5}$
11. $\frac{73}{8} = 9\frac{1}{8}$
13. $\frac{24}{6} = 4$
15. $\frac{9}{1} = 9$
17. $4\frac{2}{3} = \frac{3(4) + 2}{3} = \frac{14}{3}$
19. $8 = \frac{8}{1}$
21. $6\frac{2}{9} = \frac{6(9) + 2}{9} = \frac{56}{9}$
23. $3\frac{3}{7} = \frac{3(7) + 3}{7} = \frac{24}{7}$
25. $7\frac{6}{13} = \frac{13(7) + 6}{13} = \frac{97}{13}$
27. $10\frac{2}{5} = \frac{5(10) + 2}{5} = \frac{52}{5}$
29. $100\frac{2}{3} = \frac{3(100) + 2}{3} = \frac{302}{3}$
31. $118\frac{3}{4} = \frac{4(118) + 3}{4} = \frac{475}{4}$

## EXERCISES 5.4

1. The cross products $(1)(5) = 5$ and $(3)(3) = 9$ are not equal. The fractions are not equivalent.
3. The cross products $(1)(28) = 28$ and $(7)(4) = 28$ are equal. The fractions are equivalent.
5. The cross products $(5)(18) = 90$ and $(6)(15) = 90$ are equal. The fractions are equivalent.
7. The cross products $(2)(25) = 50$ and $(21)(4) = 84$ are not equal. The fractions are not equivalent.

9. The cross products (2)(11) = 22 and (7)(3) = 21 are not equal. The fractions are not equivalent.

11. The cross products (16)(60) = 960 and (24)(40) = 960 are equal. The fractions are equivalent.

EXERCISES 5.5

1. $\frac{8}{12} = \frac{2\cdot2\cdot2}{2\cdot2\cdot3} = \frac{2}{3}$

3. $\frac{10}{14} = \frac{2\cdot5}{2\cdot7} = \frac{5}{7}$

5. $\frac{12}{18} = \frac{2\cdot2\cdot3}{2\cdot3\cdot3} = \frac{2}{3}$

7. $\frac{35}{40} = \frac{5\cdot7}{5\cdot8} = \frac{7}{8}$

9. $\frac{11}{44} = \frac{11}{4\cdot11} = \frac{1}{4}$

11. $\frac{12}{36} = \frac{12}{12\cdot3} = \frac{1}{3}$

13. $\frac{24}{27} = \frac{3\cdot8}{3\cdot9} = \frac{8}{9}$

15. $\frac{32}{40} = \frac{8\cdot4}{8\cdot5} = \frac{4}{5}$

17. $\frac{75}{105} = \frac{15\cdot5}{15\cdot7} = \frac{5}{7}$

19. $\frac{48}{60} = \frac{12\cdot4}{12\cdot5} = \frac{4}{5}$

21. $\frac{105}{135} = \frac{15\cdot7}{15\cdot9} = \frac{7}{9}$

23. $\frac{66}{110} = \frac{22\cdot3}{22\cdot5} = \frac{3}{5}$

25. $\frac{16}{21} = \frac{2\cdot2\cdot2\cdot2}{3\cdot7} = \frac{16}{21}$, no common factors; already in simplest form.

27. $\frac{31}{52} = \frac{31}{2\cdot2\cdot13} = \frac{31}{52}$, already in simplest form.

29. $\frac{25}{100} = \frac{25}{4\cdot25} = \frac{1}{4}$ of a dollar

31. $\frac{15}{60} = \frac{15}{4\cdot15} = \frac{1}{4}$ of an hour

33. $\frac{70}{100} = \frac{10\cdot7}{10\cdot10} = \frac{7}{10}$ of a meter

35. $\frac{175}{250} = \frac{25\cdot7}{25\cdot10} = \frac{7}{10}$ of the landfill has been used

37. $\frac{15}{200} = \frac{5\cdot3}{5\cdot40} = \frac{3}{40}$ being repaired

EXERCISES 5.6

1. $\frac{1}{2} = \frac{?}{8}$ since $2\cdot4 = 8$, $? = 1\cdot4$
$? = 4$

3. $\frac{3}{7} = \frac{?}{21}$ since $7\cdot3 = 21$, $? = 3\cdot3$
$? = 9$

5. $\frac{2}{5} = \frac{?}{60}$ since $5\cdot12 = 60$, $? = 2\cdot12$
$? = 24$

7. $\frac{2}{7} = \frac{?}{35}$ since $7\cdot5 = 35$, $? = 2\cdot5$
$? = 10$

9. $\frac{7}{11} = \frac{?}{99}$ since $11\cdot9 = 99$, $? = 7\cdot9$
$? = 63$

11. $\frac{3}{8} = \frac{?}{32}$ since $8\cdot4 = 32$, $? = 3\cdot4$
$? = 12$

13. $\frac{7}{9} = \frac{?}{108}$ since $9\cdot12 = 108$, $? = 7\cdot12$
$? = 84$

15. $\frac{3}{10} = \frac{?}{200}$ since $10\cdot20 = 200$, $? = 3\cdot20$
$? = 60$

17. $\frac{12}{17} = \frac{12}{17}\cdot\frac{10}{10} = \frac{120}{170}$; $\frac{9}{10}\cdot\frac{17}{17} = \frac{153}{170}$. From smallest to largest, $\frac{12}{17}$, $\frac{9}{10}$

19. $\frac{3}{5}\cdot\frac{8}{8} = \frac{24}{40}$; $\frac{5}{8}\cdot\frac{5}{5} = \frac{25}{40}$. From smallest to largest, $\frac{3}{5}$, $\frac{5}{8}$

21. $\frac{3}{8}\cdot\frac{3}{3} = \frac{9}{24}$; $\frac{1}{3}\cdot\frac{8}{8} = \frac{8}{24}$; $\frac{1}{4}\cdot\frac{6}{6} = \frac{6}{24}$ From smallest to largest, $\frac{1}{4}$, $\frac{1}{3}$, $\frac{3}{8}$

23. $\frac{11}{12}\cdot\frac{5}{5} = \frac{55}{60}$; $\frac{4}{5}\cdot\frac{12}{12} = \frac{48}{60}$; $\frac{5}{6}\cdot\frac{10}{10} = \frac{50}{60}$ From smallest to largest, $\frac{4}{5}$, $\frac{5}{6}$, $\frac{11}{12}$

25. $\frac{5}{6}\cdot\frac{5}{5} = \frac{25}{30}$; $\frac{2}{5}\cdot\frac{6}{6} = \frac{12}{30}$; $\frac{5}{6} > \frac{2}{5}$

27. $\frac{4}{9}\cdot\frac{7}{7} = \frac{28}{63}$; $\frac{3}{7}\cdot\frac{9}{9} = \frac{27}{63}$; $\frac{4}{9} > \frac{3}{7}$

29. $\frac{9}{20}\cdot\frac{5}{5} = \frac{35}{100}$; $\frac{9}{25}\cdot\frac{4}{4} = \frac{36}{100}$; $\frac{7}{20} > \frac{9}{25}$

31. $\frac{5}{16}\cdot\frac{5}{5} = \frac{25}{80}$; $\frac{7}{20}\cdot\frac{4}{4} = \frac{28}{80}$; $\frac{5}{16} < \frac{7}{20}$

33. $\frac{2}{5} = \frac{2}{5}\cdot\frac{4}{4} = \frac{8}{20}$; $\frac{1}{4} = \frac{1}{4}\cdot\frac{5}{5} = \frac{5}{20}$

35. $\frac{5}{8} = \frac{5}{8}\cdot\frac{3}{3} = \frac{15}{24}$; $\frac{5}{12} = \frac{5}{12}\cdot\frac{2}{2} = \frac{10}{24}$

37. $\frac{1}{2} = \frac{1}{2}\cdot\frac{6}{6} = \frac{6}{12}$; $\frac{1}{3} = \frac{1}{3}\cdot\frac{4}{4} = \frac{4}{12}$; $\frac{1}{4} = \frac{1}{4}\cdot\frac{3}{3} = \frac{3}{12}$

39. $\frac{2}{15} = \frac{2}{15}\cdot\frac{7}{7} = \frac{14}{105}$; $\frac{5}{7} = \frac{5}{7}\cdot\frac{15}{15} = \frac{75}{105}$; $\frac{3}{5} = \frac{3}{5}\cdot\frac{21}{21} = \frac{63}{105}$

41. $\frac{3}{8}\cdot\frac{4}{4} = \frac{12}{32}$; $\frac{5}{16}\cdot\frac{2}{2} = \frac{10}{32}$; $\frac{11}{32}$. The $\frac{3}{8}$ drill bit is largest.

43. $\frac{3}{4}\cdot\frac{2}{2} = \frac{6}{8}$; $\frac{1}{2}\cdot\frac{4}{4} = \frac{4}{8}$. The $\frac{3}{4}$ in. plywood is thickest.

EXERCISES 5.6

## CHAPTER 6
## MULTIPLICATION AND DIVISION OF FRACTIONS

### EXERCISES 6.1

1. $\frac{3}{4}\cdot\frac{5}{11} = \frac{3\cdot 5}{4\cdot 11} = \frac{15}{44}$

3. $\frac{3}{4}\cdot\frac{7}{5} = \frac{3\cdot 7}{4\cdot 5} = \frac{21}{20} = 1\frac{1}{20}$

5. $\frac{3}{5}\cdot\frac{5}{7} = \frac{3\cdot 5}{5\cdot 7} = \frac{3}{7}$

7. $\frac{6}{13}\cdot\frac{4}{9} = \frac{6\cdot 4}{13\cdot 9} = \frac{24}{117} = \frac{8}{39}$

9. $\frac{6}{11}\cdot\frac{7}{18} = \frac{6\cdot 7}{11\cdot 18} = \frac{7}{11\cdot 3} = \frac{7}{33}$

11. $\frac{3}{10}\cdot\frac{5}{9} = \frac{3\cdot 5}{10\cdot 9} = \frac{1}{2\cdot 3} = \frac{1}{6}$

13. $\frac{7}{9}\cdot\frac{6}{5} = \frac{7\cdot 6}{9\cdot 5} = \frac{7\cdot 2}{3\cdot 5} = \frac{14}{15}$

15. $\frac{3}{4}$ of $\frac{6}{7} = \frac{3}{4}\cdot\frac{6}{7} = \frac{3\cdot 6}{4\cdot 7} = \frac{3\cdot 3}{2\cdot 7} = \frac{9}{14}$

17. $\frac{2}{7}$ of $\frac{7}{9} = \frac{2}{7}\cdot\frac{7}{9} = \frac{2\cdot 7}{7\cdot 9} = \frac{2}{9}$

### EXERCISES 6.2

1. $1\frac{2}{5}\cdot\frac{3}{4} = \frac{7}{5}\cdot\frac{3}{4} = \frac{21}{20} = 1\frac{1}{20}$

3. $\frac{5}{8}\cdot 1\frac{3}{4} = \frac{5}{8}\cdot\frac{7}{4} = \frac{5\cdot 7}{8\cdot 4} = \frac{35}{32} = 1\frac{3}{32}$

5. $3\frac{1}{3}\cdot\frac{9}{11} = \frac{10}{3}\cdot\frac{9}{11} = \frac{10\cdot 9}{3\cdot 11} = \frac{10\cdot 3}{11} = \frac{30}{11} = 2\frac{8}{11}$

7. $3\frac{1}{3}\cdot\frac{3}{7} = \frac{10}{3}\cdot\frac{3}{7} = \frac{10\cdot 3}{3\cdot 7} = \frac{10}{7} = 1\frac{3}{7}$

9. $2\frac{1}{3}\cdot 2\frac{1}{6} = \frac{7}{3}\cdot\frac{13}{6} = \frac{91}{18} = 5\frac{1}{18}$

11. $1\frac{3}{4}\cdot 2\frac{3}{5} = \frac{7}{4}\cdot\frac{13}{5} = \frac{7\cdot 13}{4\cdot 5} = \frac{91}{20} = 4\frac{11}{20}$

13. $3\frac{2}{5}\cdot 1\frac{2}{3} = \frac{17}{5}\cdot\frac{5}{3} = \frac{17}{3} = 5\frac{2}{3}$

15. $5\cdot\frac{4}{7} = \frac{5\cdot 4}{7} = \frac{20}{7} = 2\frac{6}{7}$

17. $\frac{3}{7}\cdot 14 = \frac{3\cdot 2\cdot 7}{7} = 6$

19. $15\cdot\frac{5}{6} = \frac{3\cdot 5\cdot 5}{2\cdot 3} = \frac{25}{2} = 12\frac{1}{2}$

21. $1\frac{2}{5}\cdot 1\frac{1}{4} = \frac{7}{5}\cdot\frac{5}{4} = \frac{7\cdot 5}{5\cdot 4} = \frac{7}{4} = 1\frac{3}{4}$

23. $2\frac{1}{2}\cdot 3 = \frac{5}{2}\cdot\frac{3}{1} = \frac{15}{2} = 7\frac{1}{2}$

25. $5\cdot 4\frac{2}{3} = \frac{5}{1}\cdot\frac{14}{3} = \frac{5\cdot 14}{1\cdot 3} = \frac{70}{3} = 23\frac{1}{3}$

27. $1\frac{4}{9}\cdot 3 = \frac{13}{9}\cdot\frac{3}{1} = \frac{13}{3} = 4\frac{1}{3}$

29. $2\frac{2}{5}\cdot 15 = \frac{12}{5}\cdot\frac{15}{1} = 12\cdot 3 = 36$

31. $\frac{2}{3}\cdot 6 = \frac{2\cdot 2\cdot 3}{3} = 4$ cups

33. $3\frac{5}{6}\cdot\frac{3}{4} = \frac{23}{6}\cdot\frac{3}{4} = \frac{23}{8} = 2\frac{7}{8}\ \text{ft}^2$

EXERCISES 6.3

1. $\frac{3}{7}\cdot\frac{1}{9} = \frac{3\cdot 1}{7\cdot 9} = \frac{1}{7\cdot 3} = \frac{1}{21}$

3. $\frac{6}{11}\cdot\frac{33}{12} = \frac{6\cdot 33}{11\cdot 12} = \frac{1\cdot 3}{1\cdot 2} = \frac{3}{3}$

5. $\frac{10}{12}\cdot\frac{16}{25} = \frac{10\cdot 16}{12\cdot 25} = \frac{2\cdot 4}{3\cdot 5} = \frac{8}{15}$

7. $\frac{21}{25}\cdot\frac{30}{7} = \frac{21\cdot 30}{25\cdot 7} = \frac{3\cdot 6}{5\cdot 1} = \frac{18}{5} = 3\frac{3}{5}$

9. $3\frac{2}{3}\cdot\frac{9}{10} = \frac{11}{3}\cdot\frac{9}{10} = \frac{11\cdot 9}{3\cdot 10} = \frac{11\cdot 3}{1\cdot 10} = \frac{33}{10} = 3\frac{3}{10}$

11. $5\frac{1}{3}\cdot\frac{7}{8} = \frac{16}{3}\cdot\frac{7}{8} = \frac{16\cdot 7}{3\cdot 8} = \frac{2\cdot 7}{3\cdot 1} = \frac{14}{3} = 4\frac{2}{3}$

13. $1\frac{1}{3}\cdot 1\frac{1}{5} = \frac{4}{3}\cdot\frac{6}{5} = \frac{6\cdot 4}{3\cdot 5} = \frac{2\cdot 4}{1\cdot 5} = \frac{8}{5} = 1\frac{3}{5}$

15. $2\frac{2}{7}\cdot 2\frac{1}{3} = \frac{16}{7}\cdot\frac{7}{3} = \frac{16\cdot 7}{7\cdot 3} = \frac{16\cdot 1}{1\cdot 3} = \frac{16}{3} = 5\frac{1}{3}$

17. $3\frac{3}{7}\cdot 2\frac{5}{8} = \frac{24}{7}\cdot\frac{21}{8} = \frac{24\cdot 21}{7\cdot 8} = \frac{3\cdot 3}{1\cdot 1} = 9$

19. $6\cdot 2\frac{1}{3} = \frac{6}{1}\cdot\frac{7}{3} = \frac{6\cdot 7}{1\cdot 3} = \frac{2\cdot 7}{1\cdot 1} = 14$

21. $4\frac{2}{7}\cdot 8 = \frac{30}{7}\cdot\frac{8}{1} = \frac{30\cdot 8}{7\cdot 1} = \frac{240}{7} = 34\frac{2}{7}$

23. $\frac{7}{12}\cdot\frac{3}{4}\cdot\frac{8}{15} = \frac{7\cdot 3\cdot 8}{12\cdot 4\cdot 15} = \frac{7\cdot 1\cdot 2}{12\cdot 1\cdot 5} = \frac{7\cdot 1\cdot 1}{6\cdot 1\cdot 5} = \frac{7}{30}$

25. $4\frac{1}{5}\cdot\frac{10}{21}\cdot\frac{9}{20} = \frac{21}{5}\cdot\frac{10}{21}\cdot\frac{9}{20} = \frac{21\cdot 10\cdot 9}{5\cdot 21\cdot 20}$

$= \frac{1\cdot 1\cdot 9}{5\cdot 1\cdot 2} = \frac{9}{10}$

27. $3\frac{1}{3}\cdot\frac{4}{5}\cdot 1\frac{1}{8} = \frac{10}{3}\cdot\frac{4}{5}\cdot\frac{9}{8} = \frac{10\cdot 4\cdot 9}{3\cdot 5\cdot 8}$

$= \frac{2\cdot 1\cdot 3}{1\cdot 1\cdot 2} = \frac{1\cdot 1\cdot 3}{1\cdot 1\cdot 1} = 3$

29. $\frac{2}{3}\cdot\frac{3}{7} = \frac{2\cdot 3}{3\cdot 7} = \frac{2\cdot 1}{1\cdot 7} = \frac{2}{7}$

31. $\frac{3}{5}$ of $15 = \frac{3}{5}\cdot\frac{15}{1} = \frac{3\cdot 15}{5\cdot 1} = \frac{3\cdot 3}{1\cdot 1} = 9$

33. $\frac{3}{4}$ of $2\frac{2}{5} = \frac{3}{4}\cdot\frac{12}{5} = \frac{3\cdot 12}{4\cdot 5} = \frac{3\cdot 3}{1\cdot 5} = \frac{9}{5} = 1\frac{4}{5}$

35. $\frac{6}{7}\cdot 2\frac{4}{5} = \frac{6}{7}\cdot\frac{14}{5} = \frac{6\cdot 14}{7\cdot 5} = \frac{6\cdot 2}{1\cdot 5} = \frac{12}{5} = 2\frac{2}{5}$

37. $\frac{3}{8}$ of $200 = \frac{3}{8}\cdot 200 = \frac{3\cdot 200}{8} = 3\cdot 25 = 75$

$\frac{3}{8}$ in. represents 75 mi

39. $80\cdot\frac{3}{4} = \frac{80\cdot 3}{4} = \frac{20\cdot 3}{1} = 60$

The stack is 60 in. high.

41. circumference $= \frac{22}{7}\cdot$diameter and for diameter = 21 in.

$= \frac{22}{7}\cdot 21 = \frac{22\cdot 21}{7} = \frac{22\cdot 3}{1} = 66$

circumference = 66 in.

43. fraction of eligible voters voting = fraction registered·fraction of registered voting

$= \frac{3}{4}\cdot\frac{5}{9} = \frac{3\cdot 5}{4\cdot 9} = \frac{1\cdot 5}{4\cdot 3} = \frac{5}{12}$

$\frac{5}{12}$ of eligible voters voted.

45. $3\frac{1}{3}\cdot 3\frac{3}{4} = \frac{10}{3}\cdot\frac{15}{4} = \frac{10\cdot 15}{3\cdot 4} = \frac{5\cdot 5}{1\cdot 2} = \frac{25}{2} = 12\frac{1}{2}$

$12\frac{1}{2}$ sq yd will cover the floor.

47. Distance = rate·time $= 540\cdot 4\frac{2}{3}$

$= \frac{540}{1}\cdot\frac{14}{3} = \frac{180\cdot 14}{1\cdot 1} = 2520$

The distance flown is 2520 mi.

49. Area = $\frac{1}{2}$ base·height

$= \frac{1}{2}\cdot 3\frac{1}{3}\cdot 2\frac{1}{5} = \frac{1}{2}\cdot\frac{10}{3}\cdot\frac{12}{5} = \frac{1\cdot 10\cdot 12}{2\cdot 3\cdot 5} = \frac{1\cdot 1\cdot 4}{1\cdot 1\cdot 1} = 4$

The area of the triangle is 4 sq in.

51. $7\frac{3}{4} \to 8$; $5\frac{1}{4} \to 5$. $7\frac{3}{4}\cdot 5\frac{1}{4}$ may be estimated by $8\cdot 5 = 40$.

53. $9\frac{2}{5} \to 9$; $3\frac{5}{6} \to 4$. $9\frac{2}{5}\cdot 3\frac{5}{6}$ may be estimated by $9\cdot 4 = 36$.

55. $\frac{1}{2}$ paper, $\frac{2}{5}$ newspaper. $\frac{1}{2}\cdot\frac{1}{5} = \frac{1}{5}$ newspaper

57. $\frac{1}{14}$ glass and metal, $\frac{7}{10}$ recycle. $\frac{1}{14}\cdot\frac{7}{10} = \frac{1}{20}$ saved

59. 15 trucks, $3\frac{2}{3}$ loads. $15\cdot 3\frac{2}{3} = \frac{15}{1}\cdot\frac{11}{3} = \frac{15\cdot 11}{1\cdot 3} = \frac{5\cdot 11}{1\cdot 1} = 55$ truckloads

## EXERCISES 6.4

1. $\frac{1}{5} \div \frac{3}{4} = \frac{1}{5}\cdot\frac{4}{3} = \frac{4}{15}$

3. $\frac{2}{5} \div \frac{3}{4} = \frac{2}{5}\cdot\frac{4}{3} = \frac{2\cdot 4}{5\cdot 3} = \frac{8}{15}$

5. $\frac{8}{9} \div \frac{4}{3} = \frac{8}{9}\cdot\frac{3}{4} = \frac{2}{3}$

7. $\frac{7}{10} \div \frac{5}{9} = \frac{7}{10}\cdot\frac{9}{5} = \frac{7\cdot 9}{10\cdot 5} = \frac{63}{50} = 1\frac{13}{50}$

9. $\frac{8}{15} \div \frac{2}{5} = \frac{8}{15}\cdot\frac{5}{2} = \frac{4}{3} = 1\frac{1}{3}$

11. $\frac{5}{27} \div \frac{25}{36} = \frac{5}{27}\cdot\frac{36}{25} = \frac{5\cdot 36}{27\cdot 25} = \frac{1\cdot 4}{3\cdot 5} = \frac{4}{15}$

13. $\frac{4}{5} \div 4 = \frac{4}{5}\cdot\frac{1}{4} = \frac{1}{5}$

15. $12 \div \frac{2}{3} = \frac{12}{1} \div \frac{2}{3} = \frac{12}{1}\cdot\frac{3}{2} = \frac{12\cdot 3}{1\cdot 2} = \frac{6\cdot 3}{1\cdot 1} = 18$

17. $\frac{12}{17} \div 6 = \frac{12}{17}\cdot\frac{1}{6} = \frac{2}{17}$

19. $3 \div \frac{5}{8} = 3\cdot\frac{8}{5} = \frac{24}{5} = 4\frac{4}{5}$

21. $4\frac{1}{2} \div 6 = \frac{9}{2} \div \frac{6}{1} = \frac{9}{2}\cdot\frac{1}{6} = \frac{9\cdot 1}{2\cdot 6} = \frac{3\cdot 1}{2\cdot 2} = \frac{3}{4}$

23. $9 \div 2\frac{1}{4} = 9 \div \frac{9}{4} = 9\cdot\frac{4}{9} = 4$

25. $15 \div 3\frac{1}{3} = \frac{15}{1} \div \frac{10}{3} = \frac{15}{1}\cdot\frac{3}{10} = \frac{15\cdot 3}{1\cdot 10} = \frac{3\cdot 3}{1\cdot 2} = \frac{9}{2} = 4\frac{1}{2}$

27. $1\frac{3}{5} \div \frac{4}{15} = \frac{8}{5} \div \frac{4}{15} = \frac{8}{5}\cdot\frac{15}{4} = \frac{8\cdot 15}{5\cdot 4} = \frac{2\cdot 3}{1\cdot 1} = 6$

29. $\frac{7}{12} \div 2\frac{1}{3} = \frac{7}{12} \div \frac{7}{3} = \frac{7}{12}\cdot\frac{3}{7} = \frac{3}{12} = \frac{1}{4}$

31. $5\frac{3}{5} \div \frac{7}{15} = \frac{28}{5} \div \frac{7}{15} = \frac{28}{5}\cdot\frac{15}{7} = \frac{28\cdot 15}{5\cdot 7} = \frac{4\cdot 3}{1\cdot 1} = 12$

33. $1\frac{1}{3} \div 1\frac{1}{7} = \frac{4}{3} \div \frac{8}{7} = \frac{4}{3}\cdot\frac{7}{8} = \frac{7}{6} = 1\frac{1}{6}$

35. $3\frac{3}{4} \div 1\frac{3}{8} = \frac{15}{4} \div \frac{11}{8} = \frac{15}{4}\cdot\frac{8}{11} = \frac{15\cdot 8}{4\cdot 11} = \frac{15\cdot 2}{1\cdot 11} = \frac{30}{11} = 2\frac{8}{11}$

37. $2\frac{1}{3} \div 1\frac{5}{9} = \frac{7}{3} \div \frac{14}{9} = \frac{7}{3}\cdot\frac{9}{14} = \frac{3}{2} = 1\frac{1}{2}$

39. $5\frac{1}{4} \div 7 = \frac{21}{4} \div 7 = \frac{21}{4}\cdot\frac{1}{7} = \frac{21\cdot 1}{4\cdot 7} = \frac{3\cdot 1}{4\cdot 1} = \frac{3}{4}$

Each piece is $\frac{3}{4}$ ft in length.

41. distance = rate·time or rate = $\frac{\text{distance}}{\text{time}}$

rate = $95 \div 1\frac{1}{4} = \frac{95}{1} \div \frac{5}{4} = \frac{95}{1}\cdot\frac{4}{5} = \frac{95\cdot 4}{1\cdot 5} = \frac{19\cdot 4}{1\cdot 1} = 76$

Her average speed was 76 mi/hr.

43. number of servings = $\frac{\text{weight}}{\text{number of lbs per serving}}$

$= \frac{3\frac{1}{4}}{\frac{1}{4}} = 3\frac{1}{4} \div \frac{1}{4} = \frac{13}{4}\cdot\frac{4}{1} = 13$

The roast will provide 13 servings.

45. number of packages = $\frac{\text{weight}}{\text{number of lbs per package}}$

$= \frac{19\frac{1}{8}}{\frac{3}{8}} = 19\frac{1}{8} \div \frac{3}{8} = \frac{153}{8}\cdot\frac{8}{3} = \frac{153}{3} = 51$

51 packages can be prepared.

47. number of sheets in stack = $\frac{\text{height of stack}}{\text{thickness of sheet}}$

$= \frac{48}{\frac{3}{4}} = 48 \div \frac{3}{4}$

$= 48\cdot\frac{4}{3} = \frac{16\cdot 4}{1} = 64$

There are 64 sheets of plywood in the stack.

49. $24\frac{1}{2} \div \frac{1}{2} = \frac{49}{2} \div \frac{1}{2} = \frac{49}{2}\cdot\frac{2}{1} = 49$ cells.

# CHAPTER 7

## ADDITION AND SUBTRACTION OF FRACTIONS

**EXERCISES 7.1**

1. $\frac{3}{5}+\frac{1}{5}=\frac{3+1}{5}$
$=\frac{4}{5}$

3. $\frac{4}{11}+\frac{6}{11}=\frac{4+6}{11}$
$=\frac{10}{11}$

5. $\frac{2}{10}+\frac{3}{10}=\frac{2+3}{10}$
$=\frac{5}{10}$
$=\frac{5}{2\cdot 5}$
$=\frac{1}{2}$

7. $\frac{3}{7}+\frac{4}{7}=\frac{3+4}{7}$
$=\frac{7}{7}=\frac{7\cdot 1}{7\cdot 1}$
$=1$

9. $\frac{9}{30}+\frac{11}{30}=\frac{9+11}{30}$
$=\frac{20}{30}$
$=\frac{2\cdot 10}{3\cdot 10}$
$=\frac{2}{3}$

11. $\frac{13}{48}+\frac{23}{48}=\frac{13+23}{48}$
$=\frac{36}{48}$
$=\frac{3\cdot 12}{4\cdot 12}$
$=\frac{3}{4}$

13. $\frac{3}{7}+\frac{6}{7}=\frac{3+6}{7}$
$=\frac{9}{7}$
$=1\frac{2}{7}$

15. $\frac{7}{10}+\frac{9}{10}=\frac{7+9}{10}$
$=\frac{16}{10}$
$=1\frac{6}{10}$
$=1\frac{3}{5}$

17. $\frac{11}{12}+\frac{10}{12}=\frac{11+10}{12}$
$=\frac{21}{12}$
$=1\frac{9}{12}$
$=1\frac{3}{4}$

19. $\frac{1}{8}+\frac{1}{8}+\frac{3}{8}=\frac{1+1+3}{8}$
$=\frac{5}{8}$

21. $\frac{1}{9} + \frac{4}{9} + \frac{5}{9} = \frac{1 + 4 + 5}{9}$

$= \frac{10}{9}$

$= 1\frac{1}{9}$

23. $\frac{3}{10} + \frac{7}{10} + \frac{5}{10} = \frac{3 + 7 + 5}{10}$

$= \frac{15}{10}$

$= 1\frac{5}{10}$

$= 1\frac{1}{2}$

25. 1 dime $= \frac{1}{10}$ of a dollar

$\frac{3}{10} + \frac{2}{10} + \frac{4}{10} = \frac{3 + 2 + 4}{10}$

$= \frac{9}{10}$

3 dimes + 2 dimes + 4 dimes is $\frac{9}{10}$ of a dollar.

27. 1 hour $= \frac{1}{24}$ of a day

$\frac{7}{24} + \frac{5}{24} + \frac{6}{24} = \frac{7 + 5 + 6}{24}$

$= \frac{18}{24}$

$= \frac{3\cdot 6}{4\cdot 6}$

$= \frac{3}{4}$

7 hours + 5 hours + 6 hours is $\frac{3}{4}$ of a day.

EXERCISES 7.2

1. $3 = 3$ LCD $= 2\cdot 2\cdot 3$
$4 = 2\cdot 2$ $= 12$

3. $4 = 2\cdot 2$ LCD $= 2\cdot 2\cdot 2$
$8 = 2\cdot 2\cdot 2$ $= 8$

5. $9 = 3\cdot 3$ LCD $= 3\cdot 3\cdot 3$
$27 = 3\cdot 3\cdot 3$ $= 27$

7. $8 = 2\cdot 2\cdot 2$ LCD $= 2\cdot 2\cdot 2\cdot 3$
$12 = 2\cdot 2\cdot 3$ $= 24$

9. $14 = 2\cdot 7$ LCD $= 2\cdot 3\cdot 7$
$21 = 3\cdot 7$ $= 42$

11. $20 = 2.2\cdot 5$ LCD $= 2\cdot 2\cdot 3\cdot 5$
$30 = 2\cdot 3\cdot 5$ $= 60$

13. $30 = 2\cdot 3\cdot 5$ LCD $= 2\cdot 3\cdot 5\cdot 5$
$50 = 2\cdot 5\cdot 5$ $= 150$

15. $48 = 2\cdot 2\cdot 2\cdot 2\cdot 3$ LCD $= 2\cdot 2\cdot 2\cdot 2\cdot 3\cdot 5$
$80 = 2\cdot 2\cdot 2\cdot 2\cdot 5$ $= 240$

17. $3 = 3$ LCD $= 2\cdot 2\cdot 3\cdot 5$
$4 = 2\cdot 2$ $= 60$
$5 = 5$

19. $8 = 2\cdot 2\cdot 2$ LCD $= 2\cdot 2\cdot 2\cdot 3\cdot 5$
$10 = 2\cdot 5$ $= 120$
$15 = 3\cdot 5$

21. $5$ LCD $= 2\cdot 5\cdot 5$
$10 = 2\cdot 5$ $= 50$
$25 = 5\cdot 5$

23. $14 = 2\cdot 7$ LCD $= 2\cdot 2\cdot 2\cdot 3\cdot 7$
$24 = 2\cdot 2\cdot 2\cdot 3$ $= 168$
$28 = 2\cdot 2\cdot 7$

EXERCISES 7.3

1. $\frac{2}{3} + \frac{1}{4} = \frac{8}{12} + \frac{3}{12}$

$= \frac{8 + 3}{12}$

$= \frac{11}{12}$

3. $\frac{1}{5} + \frac{3}{10} = \frac{2}{10} + \frac{3}{10}$

$= \frac{2 + 3}{10}$

$= \frac{5}{10}$

$= \frac{1}{2}$

5. $\frac{3}{4} + \frac{1}{8} = \frac{6}{8} + \frac{1}{8}$
$= \frac{6 + 1}{8}$
$= \frac{7}{8}$

7. $\frac{1}{7} + \frac{3}{5} = \frac{5}{35} + \frac{21}{35}$
$= \frac{5 + 21}{35}$
$= \frac{26}{35}$

9. $\frac{3}{7} + \frac{3}{14} = \frac{6}{14} + \frac{3}{14}$
$= \frac{6 + 3}{14}$
$= \frac{9}{14}$

11. $\frac{7}{15} + \frac{2}{35} = \frac{49}{105} + \frac{6}{105}$
$= \frac{49 + 6}{105}$
$= \frac{55}{105}$
$= \frac{5 \cdot 11}{5 \cdot 21}$
$= \frac{11}{21}$

13. $\frac{5}{8} + \frac{1}{12} = \frac{15}{24} + \frac{2}{24}$
$= \frac{15 + 2}{24}$
$= \frac{17}{24}$

15. $\frac{5}{6} + \frac{7}{9} = \frac{15}{18} + \frac{14}{18}$
$= \frac{15 + 14}{18}$
$= \frac{29}{18}$
$= 1\frac{11}{18}$

17. $\frac{5}{12} + \frac{7}{18} = \frac{15}{36} + \frac{14}{36}$
$= \frac{15 + 14}{36}$
$= \frac{29}{36}$

19. $\frac{13}{16} + \frac{17}{24} = \frac{39}{48} + \frac{34}{48}$
$= \frac{39 + 34}{48}$
$= \frac{73}{48}$
$= 1\frac{25}{48}$

21. $\frac{1}{5} + \frac{1}{3} + \frac{1}{4} = \frac{12}{60} + \frac{20}{60} + \frac{15}{60}$
$= \frac{12 + 20 + 15}{60}$
$= \frac{47}{60}$

23. $\frac{1}{5} + \frac{7}{10} + \frac{4}{15} = \frac{6}{30} + \frac{21}{30} + \frac{8}{30}$
$= \frac{6 + 21 + 8}{30}$
$= \frac{35}{30}$
$= 1\frac{1}{6}$

25. $\frac{1}{9} + \frac{7}{12} + \frac{5}{8} = \frac{8}{72} + \frac{42}{72} + \frac{45}{72}$
$= \frac{8 + 42 + 45}{72}$
$= \frac{95}{72}$
$= 1\frac{23}{72}$

27. $\frac{5}{12} + \frac{2}{21} + \frac{11}{28} = \frac{35}{84} + \frac{8}{84} + \frac{33}{84}$
$= \frac{35 + 8 + 33}{84}$
$= \frac{76}{84} = \frac{4 \cdot 19}{4 \cdot 21}$
$= \frac{19}{21}$

29. $\frac{3}{4} + \frac{3}{8} = \frac{6}{8} + \frac{3}{8} = \frac{9}{8} = 1\frac{1}{8}$. $1\frac{1}{8}$ in. thick

31. $\frac{3}{8} + \frac{1}{3} = \frac{9}{24} + \frac{8}{24} = \frac{9 + 8}{24} = \frac{17}{24}$. $\frac{17}{24}$ is used up. $\frac{7}{24}$ remains.

33. $\frac{1}{2} + \frac{3}{4} + \frac{5}{8} = \frac{4}{8} + \frac{6}{8} + \frac{5}{8} = \frac{4 + 6 + 5}{8} = \frac{15}{8} = 1\frac{7}{8}$. $1\frac{7}{8}$ in.

35. $\frac{1}{8} + \frac{1}{20} + \frac{1}{20} + \frac{1}{40} = \frac{5}{40} + \frac{2}{40} + \frac{2}{40} + \frac{1}{40} = \frac{5 + 2 + 2 + 1}{40} = \frac{10}{40} = \frac{1}{4}$

$\frac{1}{4}$ is deducted.

## EXERCISES 7.4

1. $\frac{3}{5} - \frac{1}{5} = \frac{3 - 1}{5}$
$= \frac{2}{5}$

3. $\frac{7}{9} - \frac{4}{9} = \frac{7 - 4}{9} = \frac{3}{9} = \frac{3}{3 \cdot 3}$
$= \frac{1}{3}$

5. $\frac{13}{20} - \frac{3}{20} = \frac{13 - 3}{20}$
$= \frac{10}{20}$
$= \frac{1}{2}$

7. $\frac{19}{24} - \frac{5}{24} = \frac{19 - 5}{24}$
$= \frac{14}{24} = \frac{7 \cdot 2}{12 \cdot 2}$
$= \frac{7}{12}$

9. $\frac{4}{5} - \frac{1}{3} = \frac{12}{15} - \frac{5}{15}$
$= \frac{12 - 5}{15}$
$= \frac{7}{15}$

11. $\frac{11}{15} - \frac{3}{5} = \frac{11}{15} - \frac{9}{15}$
$= \frac{11 - 9}{15}$
$= \frac{2}{15}$

13. $\frac{3}{8} - \frac{1}{4} = \frac{3}{8} - \frac{2}{8}$
$= \frac{3 - 2}{8}$
$= \frac{1}{8}$

15. $\frac{5}{12} - \frac{3}{8} = \frac{10}{24} - \frac{9}{24}$
$= \frac{10 - 9}{24}$
$= \frac{1}{24}$

17. $\frac{13}{25} - \frac{2}{15} = \frac{39}{75} - \frac{10}{75}$
$= \frac{39 - 10}{75}$
$= \frac{29}{75}$

19. $\frac{15}{27} - \frac{7}{18} = \frac{30}{54} - \frac{21}{54}$
$= \frac{30 - 21}{54}$
$= \frac{9}{54} = \frac{1}{6}$

21. $\frac{13}{18} - \frac{7}{12} = \frac{26}{36} - \frac{21}{36}$
$= \frac{26 - 21}{36}$
$= \frac{5}{36}$

23. $\frac{33}{40} - \frac{7}{24} = \frac{99}{120} - \frac{35}{120}$
$= \frac{99 - 35}{120}$
$= \frac{64}{120} = \frac{8 \cdot 8}{8 \cdot 15}$
$= \frac{8}{15}$

25. $\frac{15}{16} - \frac{5}{8} - \frac{1}{4} = \frac{15}{16} - \frac{10}{16} - \frac{4}{16}$
$= \frac{15 - 10 - 4}{16}$
$= \frac{1}{16}$

27. $\frac{5}{9} + \frac{7}{12} - \frac{5}{8} = \frac{40}{72} + \frac{42}{72} - \frac{45}{72}$
$= \frac{40 + 42 - 45}{72}$
$= \frac{37}{72}$

29. $\frac{17}{32} - \frac{1}{4} = \frac{17}{32} - \frac{8}{32} = \frac{17 - 8}{32} = \frac{9}{32}$. The missing dimension is $\frac{9}{32}$ in.

31. $\frac{7}{8} - \frac{1}{3} = \frac{21}{24} - \frac{8}{24} = \frac{21 - 8}{24} = \frac{13}{24}$. $\frac{13}{24}$ of an acre remains.

33. $\frac{3}{4} - \frac{5}{8} = \frac{6}{8} - \frac{5}{8} = \frac{6 - 5}{8} = \frac{1}{8}$. $\frac{1}{8}$ is less than $\frac{1}{4}\left(= \frac{2}{8}\right)$ so there is not enough for the pie crust.

## EXERCISES 7.5

1. $2\frac{2}{9} + 3\frac{5}{9} = \frac{20}{9} + \frac{32}{9}$
$= \frac{20 + 32}{9} = \frac{52}{9}$
$= 5\frac{7}{9}$

3. $2\frac{1}{9} + 5\frac{5}{9} = (2 + 5) + \left(\frac{1}{9} + \frac{5}{9}\right)$
$= 7 + \frac{1 + 5}{9} = 7\frac{6}{9}$
$= 7\frac{2}{3}$

5. $6\frac{5}{9} + 4\frac{7}{9} = (6 + 4) + \left(\frac{5}{9} + \frac{7}{9}\right)$
$= 10 + \frac{5 + 7}{9}$
$= 10 + \frac{12}{9} = 10 + \frac{9}{9} + \frac{3}{9}$
$= 10 + 1 + \frac{3}{9}$
$= 11\frac{3}{9} = 11\frac{1}{3}$

7. $1\frac{1}{3} + 2\frac{1}{5} = (1 + 2) + \left(\frac{1}{3} + \frac{1}{5}\right)$
$= 3 + \frac{5}{15} + \frac{3}{15}$
$= 3 + \frac{5 + 3}{15}$
$= 3 + \frac{8}{15}$
$= 3\frac{8}{15}$

9. $5\frac{3}{8} + 3\frac{5}{12} = \frac{43}{8} + \frac{41}{12}$

$= \frac{129}{24} + \frac{82}{24}$

$= \frac{129 + 82}{24}$

$= \frac{211}{24}$

$= 8\frac{19}{24}$

11. $3\frac{5}{6} + 2\frac{3}{4} = \frac{23}{6} + \frac{11}{4}$

$= \frac{46}{12} + \frac{33}{12}$

$= \frac{46 + 33}{12}$

$= \frac{79}{12}$

$= 6\frac{7}{12}$

13. $2\frac{1}{4} + 3\frac{5}{8} + 1\frac{1}{6} = (2 + 3 + 1) + \left(\frac{1}{4} + \frac{5}{8} + \frac{1}{6}\right)$

$= 6 + \left(\frac{6}{24} + \frac{15}{24} + \frac{4}{24}\right)$

$= 6 + \frac{6 + 15 + 4}{24}$

$= 6 + \frac{25}{24}$

$= 6 + \frac{24}{24} + \frac{1}{24}$

$= 6 + 1 + \frac{1}{24}$

$= 7 + \frac{1}{24}$

$= 7\frac{1}{24}$

15. $3\frac{3}{5} + 4\frac{1}{4} + 5\frac{3}{10} = (3 + 4 + 5) + \left(\frac{3}{5} + \frac{1}{4} + \frac{3}{10}\right)$

$= 12 + \left(\frac{12}{20} + \frac{5}{20} + \frac{6}{20}\right)$

$= 12 + \frac{12 + 5 + 6}{20}$

$= 12 + \frac{23}{20}$

$= 12 + \frac{20}{20} + \frac{3}{20}$

$= 12 + 1 + \frac{3}{20}$

$= 13 + \frac{3}{20}$

$= 13\frac{3}{20}$

17. $5\frac{5}{6} + 3\frac{4}{5} + 7\frac{2}{3} = (5 + 3 + 7) + \left(\frac{5}{6} + \frac{4}{5} + \frac{2}{3}\right)$

$= 15 + \left(\frac{25}{30} + \frac{24}{30} + \frac{20}{30}\right)$

$= 15 + \frac{25 + 24 + 20}{30}$

$= 15 + \frac{69}{30}$

$= 15 + \frac{60}{30} + \frac{9}{30}$

$= 15 + 2 + \frac{9}{30}$

$= 17 + \frac{9}{30}$

$= 17\frac{9}{30}$

$= 17\frac{3}{10}$

19. $7\frac{7}{8} - 3\frac{3}{8} = (7 - 3) + \left(\frac{7}{8} - \frac{3}{8}\right)$

$= 4 + \frac{7 - 3}{8}$

$= 4 + \frac{4}{8}$

$= 4 + \frac{1}{2}$

$= 4\frac{1}{2}$

21. $3\frac{2}{5} - 1\frac{4}{5} = 2\frac{7}{5} - 1\frac{4}{5}$

$= (2 - 1) + \left(\frac{7}{5} - \frac{4}{5}\right)$

$= 1 + \frac{7 - 4}{5}$

$= 1 + \frac{3}{5}$

$= 1\frac{3}{5}$

23. $3\frac{2}{3} - 2\frac{1}{4} = (3 - 2) + \left(\frac{2}{3} - \frac{1}{4}\right)$

$= 1 + \frac{8}{12} - \frac{3}{12}$

$= 1 + \frac{8 - 3}{12}$

$= 1 + \frac{5}{12}$

$= 1\frac{5}{12}$

25. $6\frac{3}{10} - 3\frac{7}{15} = 5\frac{13}{10} - 3\frac{7}{15}$

$= (5 - 3) + \left(\frac{13}{10} - \frac{7}{15}\right)$

$= 2 + \frac{39}{30} - \frac{14}{30} = 2 + \frac{39 - 14}{30}$

$= 2 + \frac{25}{30} = 2 + \frac{5}{6}$

$= 2\frac{5}{6}$

27. $7\frac{5}{12} - 3\frac{11}{18} = 6\frac{17}{12} - 3\frac{11}{18} = (6 - 3) + \left(\frac{17}{12} - \frac{11}{18}\right)$

$= 3 + \frac{51}{36} - \frac{22}{36} = 3 + \frac{51 - 22}{36}$

$= 3 + \frac{29}{36} = 3\frac{29}{36}$

29. $5 - 2\frac{1}{4} = 4\frac{4}{4} - 2\frac{1}{4} = (4 - 2) + \left(\frac{4}{4} - \frac{1}{4}\right)$

$= 2 + \frac{4 - 1}{4} = 2 + \frac{3}{4}$

$= 2\frac{3}{4}$

31. $7 - 5\frac{4}{9} = 6\frac{9}{9} - 5\frac{4}{9} = (6 - 5) + \left(\frac{9}{9} - \frac{4}{9}\right)$

$= 1 + \frac{9 - 4}{9} = 1 + \frac{5}{9}$

$= 1\frac{5}{9}$

33. $3\frac{3}{4} + 5\frac{1}{2} - 2\frac{3}{8} = (3 + 5 - 2) + \left(\frac{3}{4} + \frac{1}{2} - \frac{3}{8}\right) = 6 + \left(\frac{6}{8} + \frac{4}{8} - \frac{3}{8}\right)$

$= 6 + \frac{6 + 4 - 3}{8} = 6 + \frac{7}{8}$

$= 6\frac{7}{8}$

35. $2\frac{3}{8} + 2\frac{1}{4} - 1\frac{5}{6} = (2 + 2) + \left(\frac{3}{8} + \frac{1}{4}\right) - 1\frac{5}{6} = 4 + \left(\frac{3}{8} + \frac{2}{8}\right) - 1\frac{5}{6}$

$= 4 + \frac{3 + 2}{8} - 1\frac{5}{6} = 4 + \frac{5}{8} - 1\frac{5}{6}$

$= 4\frac{5}{8} - 1\frac{5}{6} = 3\frac{13}{8} - 1\frac{5}{6} = (3 - 1) + \left(\frac{13}{8} - \frac{5}{6}\right)$

$= 2 + \left(\frac{39}{24} - \frac{20}{24}\right) = 2 + \left(\frac{39 - 20}{24}\right)$

$= 2 + \frac{19}{24}$

$= 2\frac{19}{24}$

## EXERCISES 7.6

1 $15\frac{5}{8} + 25\frac{3}{4} = (15 + 25) + \left(\frac{5}{8} + \frac{3}{4}\right) = 40 + \frac{5}{8} + \frac{6}{8}$

$= 40\frac{11}{8} = 41\frac{3}{8}$

The total length of pipe needed is $41\frac{3}{8}$ in.

3. $2\frac{1}{4} + 5\frac{3}{4} + 4\frac{1}{2} = (2 + 5 + 4) + \left(\frac{1}{4} + \frac{3}{4} + \frac{1}{2}\right) = 11 + \left(\frac{1}{4} + \frac{3}{4} + \frac{2}{4}\right)$

$= 11 + \frac{1 + 3 + 2}{4} = 11 + \frac{6}{4}$

$= 11 + \frac{4}{4} + \frac{2}{4} = 11 + 1 + \frac{1}{2}$

$= 12 + \frac{1}{2}$

$= 12\frac{1}{2}$

Franklin worked a total of $12\frac{1}{2}$ hours.

5. $1\frac{3}{8} + 1\frac{1}{4} + 1\frac{5}{8} = (1 + 1 + 1) + \left(\frac{3}{8} + \frac{1}{4} + \frac{5}{8}\right) = 3 + \left(\frac{3}{8} + \frac{2}{8} + \frac{5}{8}\right)$

$$= 3 + \frac{3 + 2 + 5}{8} = 3 + \frac{10}{8}$$

$$= 3 + \frac{8}{8} + \frac{2}{8} = 3 + 1 + \frac{1}{4}$$

$$= 4 + \frac{1}{4}$$

$$= 4\frac{1}{4}$$

The perimeter is $4\frac{1}{4}$ inches.

7. $\frac{3}{4} + 1\frac{1}{4} + \frac{5}{8} + \frac{1}{8} = \frac{3}{4} + \frac{5}{4} + \frac{5}{8} + \frac{1}{8} = \frac{6}{8} + \frac{10}{8} + \frac{5}{8} + \frac{1}{8}$

$$= \frac{6 + 10 + 5 + 1}{8} = \frac{22}{8}$$

$$= 2\frac{6}{8} = 2\frac{3}{4}$$

Mrs. Selden should buy $2\frac{3}{4}$ yd of fabric.

9. drop = opening − closing = $34\frac{3}{8} - 28\frac{3}{4} = (34 - 28) + \left(\frac{3}{8} - \frac{3}{4}\right)$

drop = $6 + \frac{3}{8} - \frac{6}{8} = 5 + \frac{8}{8} + \frac{3}{8} - \frac{6}{8} = 5 + \frac{11}{8} - \frac{6}{8} = 5 + \frac{11 - 6}{8} = 5 + \frac{5}{8}$

drop = $5\frac{5}{8}$

11. increase = Sept. rate − May rate = $14\frac{1}{4} - 12\frac{3}{8} = 13\frac{5}{4} - 12\frac{3}{8}$

increase = $(13 - 12) + \left(\frac{5}{4} - \frac{3}{8}\right) = 1 + \left(\frac{10}{8} - \frac{3}{8}\right) = 1 + \frac{10 - 3}{8}$

increase = $1 + \frac{7}{8}$

increase = $1\frac{7}{8}$ percent

13. dimension = $5\frac{1}{4} - 3\frac{3}{8} = 4\frac{5}{4} - 3\frac{3}{8} = (4 - 3) + \left(\frac{5}{4} - \frac{3}{8}\right)$

dimension = $1 + \frac{10}{8} - \frac{3}{8} = 1 + \frac{10 - 3}{8} = 1 + \frac{7}{8}$

dimension = $1\frac{7}{8}$ inches

15. additional hours $= 20 - 5\frac{1}{2} - 3\frac{3}{4} = \frac{40}{2} - \frac{11}{2} - 3\frac{3}{4}$

$= \frac{40 - 11}{2} - 3\frac{3}{4} = \frac{29}{2} - \frac{15}{4}$

$= \frac{58}{4} - \frac{15}{4} = \frac{58 - 15}{4}$

$= \frac{43}{4}$

$= 10\frac{3}{4}$ hours

Ben can work $10\frac{3}{4}$ hours during the week.

17. carpet used $= 20\frac{3}{4} + 15\frac{1}{2} + 6\frac{1}{4} = (20 + 15 + 6) + \left(\frac{3}{4} + \frac{2}{4} + \frac{1}{4}\right)$

carpet used $= 41 + \frac{6}{4} = 41 + \frac{3}{2} = 41 + 1 + \frac{1}{2} = 42\frac{1}{2}$

carpet used $= 42\frac{1}{2}$

carpet remaining $= 50 -$ carpet used $= 50 - 42\frac{1}{2} = 49\frac{2}{2} - 42\frac{1}{2}$

carpet remaining $= (49 - 42) + \left(\frac{2}{2} - \frac{1}{2}\right) = 7 + \frac{2 - 1}{2} = 7 + \frac{1}{2}$

carpet remaining $= 7\frac{1}{2}$ square yards.

There are $7\frac{1}{2}$ square yards remaining.

19. $8 - 2\frac{3}{4} - 2\frac{1}{2} = \frac{32}{4} - \frac{11}{4} - 2\frac{1}{2} = \frac{32 - 11}{4} - 2\frac{1}{2} = \frac{21}{4} - \frac{5}{2}$

$= \frac{21}{4} - \frac{10}{4} = \frac{21 - 10}{4} = \frac{11}{4} = 2\frac{3}{4}$

$2\frac{3}{4}$ hours remain to be driven.

21. $\frac{1}{2} + \frac{1}{10} + \frac{1}{5} = \frac{5}{10} + \frac{1}{10} + \frac{2}{10}$

$= \frac{5 + 1 + 2}{10}$

$= \frac{8}{10} = \frac{4}{5}$

Paper plastic, and diapers take up $\frac{4}{5}$

of landfills.

23. $\frac{1}{2} + \frac{1}{10} = \frac{5}{10} + \frac{1}{10} = \frac{6}{10} = \frac{3}{5}$ used by paper and plastic

$1 - \frac{3}{5} = \frac{5}{5} - \frac{3}{5} = \frac{2}{5}$

$\frac{2}{5}$ of landfill is available for materials other than paper and plastic.

# CHAPTER 8

# ADDITION, SUBTRACTION, AND MULTIPLICATION OF DECIMALS

## EXERCISES 8.1

1. The place value of 7 in 8.57932 is hundredths.
3. The place value of 3 in 8.57932 is ten–thousandths.
5. $\frac{23}{100} = 0.23$
7. $\frac{209}{10,000} = 0.0209$
9. $23\frac{56}{1000} = 23.056$
11. 0.23: twenth–three hundredths
13. 0.071: seventy–one thousandths
15. 12.07: twelve and seven hundredths
17. Fifty–one thousandths: 0.051
19. Seven and three tenths: 7.3
21. $0.65 = \frac{65}{100} = \frac{13}{20}$
23. $5.231 = 5\frac{231}{1000}$
25. $0.69 > 0.689$
27. $1.23 = 1.230$
29. $10 > 9.9$
31. $1.459 < 1.46$

## EXERCISES 8.2

1. 53.48 to tenths: 53.5
3. 21.534 to hundredths: 21.53
5. 0.342 to hundredths: 0.34
7. 2.71828 to thousandths: 2.718
9. 0.0475 to tenths: 0.0
11. 4.85344 to ten–thousandths: 4.8534
13. 6.734 to two decimal places: 6.73
15. 6.58739 to four decimal places: 6.5874
17. 56.35829: 56.4 to tenths
19. 56.35829: 56.358 to thousandths

## EXERCISES 8.3

1.
$$\begin{array}{r} 0.28 \\ +\ 0.79 \\ \hline 1.07 \end{array}$$

3.
$$\begin{array}{r} 1.045 \\ +\ 0.23 \\ \hline 1.275 \end{array}$$

5.
$$\begin{array}{r} 0.62 \\ 4.23 \\ +\ 12.5 \\ \hline 17.35 \end{array}$$

7.
$$\begin{array}{r} 5.28 \\ +\ 19.455 \\ \hline 24.735 \end{array}$$

9.
$$\begin{array}{r} 13.58 \\ 7.239 \\ +\ 1.5 \\ \hline 22.319 \end{array}$$

11.
$$\begin{array}{r} 25.3582 \\ 6.5 \\ 1.898 \\ +\ 0.69 \\ \hline 34.4462 \end{array}$$

13. $0.43 + 0.8 + 0.561 = 1.791$
15. $5 + 23.7 + 8.7 + 9.85 = 47.25$
17. $25.83 + 1.7 + 3.92 = 31.45$
19. $42.731 + 1.058 + 103.24 = 147.029$
21. $0.23 + 0.5 + 0.268 = 0.998$
23. $5.3 + 0.75 + 20.13 + 12.7 = 38.88$

25. 12.7 + 15.9 + 13.8 = 42.4, Dien bought 42.4 gallons of gas on the three day trip.

27. 5.38 + 3.2 + 4.79 = 13.37, the rainfull during the winter months was 13.37 cm.

29. 45.69 + 123 + 95.60 + 8.65 = 272.94, Anna's total expenses during the trip were $272.94.

31. 50 + 11.38 + 112.57 + 9.73 = 183.68, Bruce wrote checks totalling $183.68 during this week.

33. 18 + 7 + 9 + 7 + 2 = $43

35. 457 + 124 + 212 + 416 = $1209

37. 0.56 + 0.5657 + 0.5780 = $1.7037 billion

39. 0.011 + 0.1215 + 0.0746 = $0.2071 billion

## EXERCISES 8.4

1. $\begin{array}{r} 0.85 \\ -\ 0.59 \\ \hline 0.26 \end{array}$

3. $\begin{array}{r} 23.81 \\ -\ 6.57 \\ \hline 17.24 \end{array}$

5. $\begin{array}{r} 17.134 \\ -\ 3.502 \\ \hline 13.632 \end{array}$

7. $\begin{array}{r} 35.8 \\ -\ 7.45 \\ \hline 28.35 \end{array}$

9. $\begin{array}{r} 3.82 \\ -\ 1.565 \\ \hline 2.255 \end{array}$

11. $\begin{array}{r} 7.32 \\ -\ 4.7 \\ \hline 2.62 \end{array}$

13. $\begin{array}{r} 12. \\ -\ 5.35 \\ \hline 6.65 \end{array}$

15. $\begin{array}{r} 15.02 \\ -\ 2.545 \\ \hline 12.475 \end{array}$

17. $\begin{array}{r} 28. \\ -\ 24.725 \\ \hline 3.275 \end{array}$

19. $\begin{array}{r} 6.84 \\ -\ 2.87 \\ \hline 3.97 \end{array}$

21. $\begin{array}{r} 9.4 \\ -\ 7.75 \\ \hline 1.65 \end{array}$

23. $\begin{array}{r} 5. \\ -\ 0.24 \\ \hline 4.76 \end{array}$

25. saving = selling price − discounted price
saving = 399.5 − 365.75 = 33.75, the saving on the set is $33.75.

27. amount above normal = temperature − normal temperature
= 101.3 − 98.6
= 2.7, the temperature of 101.3 is 2.7°F above normal.

29. 0.65 + 0.375 + a = 2.000
1.025 + a = 2.000
a = 2.000 − 1.025
a = 0.975, the missing dimension is 0.975 in.

31. amount driven = final odometer reading − initial odometer reading
= 16479.8 − 15785.3 = 694.5, on her trip Laura drove 694.5 mi.

33.

| 1.6 | a | 1.2 |
|---|---|---|
| b | 1 | c |
| 0.8 | d | e |

adding diagonal gives 0.8 + 1 + 1.2 = 3. The first row must add to 3, so 1.6 + a + 1.2 = 3, from which a = 0.2.
From other diagonal 1.6 + 1 + e = 3 from which e = 0.4.
From third column 1.2 + c + 0.4 = 3 from which c = 1.4.
From second row b + 1 + 1.4 = 3 from which b = 0.6.
From third row 0.8 + d + 0.4 = 3 from which d = 1.8.
As a check note that all rows, columns, and diagonals add to 3.

The final square is

| 1.6 | 0.2 | 1.2 |
|---|---|---|
| 0.6 | 1 | 1.4 |
| 0.8 | 1.8 | 0.4 |

35. $$\begin{array}{r} 18.110 \\ -\ 5.796 \\ \hline 12.314 \end{array}$$ billion liters per day

37. $$\begin{array}{r} 0.113 \\ -\ 0.058 \\ \hline 0.055 \end{array}$$ more used in the bathroom sink

39. $(0.330 + 0.196 + 0.113) = 0.639$
$(0.025 + 0.058 + 0.018) = 0.101$
$0.639 - 0.101 = 0.538$ more used in the American bathroom

## EXERCISES 8.5

1. $$\begin{array}{r} 2.3 \\ \times\ 3.4 \\ \hline 92 \\ 69\phantom{0} \\ \hline 7.82 \end{array}$$

3. $$\begin{array}{r} 8.4 \\ \times\ 5.2 \\ \hline 168 \\ 420\phantom{0} \\ \hline 43.68 \end{array}$$

5. $$\begin{array}{r} 2.56 \\ \times\ \ 72 \\ \hline 512 \\ 1792\phantom{0} \\ \hline 184.32 \end{array}$$

7. $$\begin{array}{r} 0.78 \\ \times\ 2.3 \\ \hline 234 \\ 156\phantom{0} \\ \hline 1.794 \end{array}$$

9. $$\begin{array}{r} 15.7 \\ \times\ 2.35 \\ \hline 785 \\ 471\phantom{0} \\ 314\phantom{00} \\ \hline 36.895 \end{array}$$

11. $$\begin{array}{r} 0.354 \\ \times\ \ 0.8 \\ \hline 0.2832 \end{array}$$

13. $$\begin{array}{r} 3.28 \\ \times\ 5.07 \\ \hline 2296 \\ 1640\phantom{00} \\ \hline 16.6296 \end{array}$$

15. $$\begin{array}{r} 5.238 \\ \times\ \ 0.48 \\ \hline 41904 \\ 20952\phantom{0} \\ \hline 2.51424 \end{array}$$

17. $$\begin{array}{r} 1.053 \\ \times\ 0.552 \\ \hline 2106 \\ 5265\phantom{0} \\ 5265\phantom{00} \\ \hline 0.581256 \end{array}$$

19. $$\begin{array}{r} 0.0056 \\ \times\ 0.082 \\ \hline 112 \\ 448\phantom{0} \\ \hline 0.0004592 \end{array}$$

21. $0.8 \times 2.376 = 1.9008$

23. $0.3085 \times 4.5 = 1.38825$

25. total cost = cost per shirt × number of shirts
= 9.98 × 4 = 39.92
Kurt spent a total of $39.92 on the four shirts.

27. total weight = weight per gallon × number of gallons
= 8.34 × 2.5 = 20.85
2.5 gallons of water weighs 20.85 lbs.

29. interest = amount of loan × 0.095
= 1500 × 0.095 = 142.50
The interest on the loan is $142.50 for 1 year.

31. circumference = diamter × 3.14
= 7.15 × 3.14 = 22.451
A circle with a diameter of 7.15 in. has a circumference of 22.451 in.

33. area = length × width
= 28 × 21.6 = 604.8
A sheet of paper 21.6 by 28 cm has an area of 604.8 $cm^2$.

35. number of centimeters = number of inches × number of cm per in
= 5.3 × 2.54 = 13.462
To two decimal places, 5.3 inches is 13.46 cm.

37. cost = cost per yard × number of yards
= 15.49 × 7.8 = 120.822
The cost of the carpet to the nearest cent is $120.82.

39. 7.9 → 8 area = 8 × 11 = 88, 88 $m^2$ is an estimate for the area
    11.2 → 11

41. 0.87 × 0.03 = 0.0261

43. 15.5 × 0.67 = 10.385 liters

**EXERCISES 8.6**

1. 5.89 × 10 = 58.9

3. 23.79 × 100 = 2379

5. 0.045 × 10 = 0.45

7. 0.431 × 100 = 43.1

9. 0.471 × 100 = 47.1

11. 0.7125 × 1000 = 712.5

13. 4.25 × $10^2$ = 425

15. 3.45 × $10^4$ = 34,500

17. total cost = cost per item × number of items
    = 1.38 × 100 = 138
    The total cost of the items is $138.00.

19. number of grams = number of kilograms × 1000
    = 2.2 × 1000 = 2200
    There are 2200 grams in 2.2 kilograms.

21. 18.25 → 18
    6.80 → 7
    8.75 → 9
    7.40 → 7
    1.70 → 2
    $43 $43 is an estimate for the bill.

# CHAPTER 9
# DIVISION OF DECIMALS

## EXERCISES 9.1

```
      2.78
1.  6)16.68        16.68 ÷ 6 = 2.78
      12
       4 6
       4 2
        48
        48
         0
```

```
      0.48
3.  4)1.98         1.92 ÷ 4 = 0.48
      1 6
        32
        32
         0
```

```
     0.685
5.  8)5.480        5.48 ÷ 8 = 0.685
      4 8
        68
        64
         40
         40
          0
```

```
      2.315
7.  6)13.890       13.89 ÷ 6 = 2.315
      12
       1 8
       1 8
          9
          6
          30
          30
           0
```

```
        5.8
9.  32)185.6       185.6 ÷ 32 = 5.8
       160
        25 6
        25 6
           0
```

```
        2.35
11. 34)79.90       79.9 ÷ 34 = 2.35
       68
       11 9
       10 2
        1 70
        1 70
           0
```

```
       0.265
13. 52)13.780      13.78 ÷ 52 = 0.265
       10 4
        3 38
        3 12
          260
          260
            0
```

```
        3.214
15. 45)144.630     144.63 ÷ 45 = 3.214
       135
         9 6
         9 0
           63
           45
           180
           180
             0
```

```
      2.64
17.  9)23.80       23.8 ÷ 9 = 2.644··· = 2.64̄
       18                    = 2.6 to the nearest tenth
        5 8
        5 4
         40
         36
          4
```

```
     0.836
19. 46)38.480
      36 8
       1 68
       1 38
         300
         276
          24
```

$38.48 \div 46 = 0.836\cdots$
$= 0.84$ to the nearest hundredth

```
      2.411
21. 52)125.400
      104
       21 4
       20 8
          60
          52
          80
          52
          28
```

$125.4 \div 52 = 2.411\cdots$
$= 2.4$ to the nearest tenth

```
     0. 33107
23. 28)0.927000
        84
         87
         84
          30
          28
          200
          196
            4
```

$0.927 \div 28 = 0.033107\cdots$
$= 0.033$ to the nearest thousandth

25. cost per record = total cost $\div$ number of records
$= 13.47 \div 3 = 4.49$
Marv's cost per record was \$4.49.

27. cost per book = total cost $\div$ number of books
$= 190.25 \div 72 = 2.642361111\cdots = 2.64236\overline{1}$
Average cost per book to nearest cent was \$2.64.

29. cost per item = total cost $\div$ number of items
$= 28.20 \div 48 = 0.5875$
The cost of an individual item to the nearest cent is 59¢.

31. monthly payment = (cost − down payment) $\div$ number of payments
$= (736.12 - 100) \div 18$
$= 636.12 \div 18 = 35.34$
Al's monthly payment is \$35.34.

33. average miles per gallon = sum of readings $\div$ number of readings
$= (32.3 + 31.6 + 29.5 + 27.3 + 33.4) \div 5$
$= 154.1 \div 5 = 30.82$
Lucia's average mileage to the nearest tenth of mile per gallon was 30.8.

35. grade-point average = grade point average per semester $\div$ number of semesters
$= (2.81 + 3.05 + 3.62 + 2.95 + 3.15 + 2.79 + 3.45 + 3.53) \div 8$
$= 25.35 \div 8 = 3.16875$

Jeremy's grade-point average, to the nearest hundredth, was 3.17.

**EXERCISES 9.2**

```
          1 8.45
1.  0.6)11.0 70
         6
         5 0
         4 8
           2 7
           2 4
             30
             30
              0
```

```
        1.9
3.  3.8)7.2 2
        3 8
        3 4 2
        3 4 2
            0
```

```
            2.235
5.  5.2)11.6 220
        10 4
         1 2 2
         1 0 4
           1 82
           1 56
             260
             260
               0
```

```
            6.85
7.  0.27)1.84 95
         1 62
           22 9
           21 6
            1 35
            1 35
               0
```

```
              34.5
9.  0.046)1.587 0
          1 38
            207
            184
             23 0
             23 0
                0
```

```
            0.235
11.  2.8)0.6 580
          5 6
            98
            84
            140
            140
              0
```

0.658 ÷ 2.8 = 0.235

```
               6.25
13.  0.524)3.275 00
           3 144
             131 0
             104 8
              26 20
              26 20
                  0
```

3.275 ÷ 0.524 = 6.25

```
           2.345
15.  0.7)1.6 420
         1 4
           2 4
           2 1
             32
             28
              40
              35
               5
```

```
       2.35 to nearest hundredth
0.7)1.6 42
```

```
            1.87                             1.9  to nearest tenth
17.  4.5)8.4 15                      4.5)8.4 15
         4 5
         3 9 1
         3 6 0
           3 15
           3 15
              0

            1.522                            1.52  to nearest hundredth
19.  3.12)4.75 000                   3.12)4.75
          3 12
          1 63 0
          1 56 0
             7 00
             6 24
              760
              624
              136

            0.0381                           0.038 to nearest thousandth
21.  5.38)0.20 5000                  5.38)0.205
            16 14
             4 360
             4 304
               560
               538
                22

            0.537                            0.54 to nearest hundredth
23.  2.42)1.30 000                   2.42)1.3
          1 21 0
             9 00
             7 26
             1 740
             1 694
                46

              1.5865
25.  0.624)0.990 0000                0.99 ÷ 0.624 = 1.587 to nearest thousandth
             624
             366 0
             312 0
              54 00
              49 92
               4 080
               3 744
                 3360
                 3120
                  240
```

```
              0.05854
27.  2.135 )0.125 00000          0.125 ÷ 2.135 = 0.0585 to nearest ten-thousandth
            106 75
            18 250
            17 080
             1 1700
             1 0675
               10250
                8540
                1710
```

29. number of labels = length of tape ÷ length of label
    = 91.25 ÷ 1.25 = 73
    73 labels 1.25 in. long can be made from 91.25 in. of tape.

31. cost per pound = total cost ÷ number of pounds
    = 14.89 ÷ 5.3 = 2.809433962···
    The cost per pound to the nearest cent is $2.81.

33. miles per gallon = total number of miles ÷ total number of gallons
    = 1390 ÷ (15.5 + 16.2 + 10.8)
    = 1390 ÷ 42.5 = 32.70588235···
    The number of miles per gallon to the nearest tenth is 32.7.

35. mpg $= \frac{19{,}315 - 18{,}912}{22.9} = \frac{403}{22.9} = 17.6$ mph to nearest tenth

37. $\frac{35}{25.4} = 1.38$ in. to nearest hundredth

39. $\frac{32.5}{2.75} = 11.82$ (12 shaves)

41. $\frac{105}{52.5} = 2$ loads

## EXERCISES 9.3

1. 5.8 ÷ 10 = 0.58

3. 4.568 ÷ 100 = 0.04568

5. 24.39 ÷ 1000 = 0.02439

7. 6.9 ÷ 1000 = 0.0069

9. $7.8 \div 10^2 = 7.8 \div 100 = 0.078$

11. $45.2 \div 10^5 = 45.2 \div 100{,}000 = 0.000452$

13. cost per homeowner = cost of project ÷ number of homeowners
    = 4850 ÷ 10 = 485
    Each homeowner will pay $485.

15. cost per fixture = total cost ÷ number of fixtures
    = 2780 ÷ 100 = 27.80
    Each fixture cost $27.80.

17. mass in grams = mass in mg ÷ 100
    = 250 ÷ 1000 = 0.250
    A 250 mg tablet has a mass of 0.250 gram.

19. cost per calculator = total cost ÷ number of calculators
    = $593.88 ÷ 100 = 5.9388
    To the nearest cent, each calculator cost is $5.94.

EXERCISES 9.4

1. $$\begin{array}{r} 0.75 \\ 4\overline{)3.00} \\ \underline{2\ 8} \\ 20 \\ \underline{20} \\ 0 \end{array}$$

$\frac{3}{4} = 0.75$

3. $$\begin{array}{r} 0.45 \\ 20\ \overline{)9.00} \\ \underline{8\ 0} \\ 1\ 00 \\ \underline{1\ 00} \\ 0 \end{array}$$

$\frac{9}{20} = 0.45$

5. $$\begin{array}{r} 0.2 \\ 5\overline{)1.0} \\ \underline{1\ 0} \\ 0 \end{array}$$

$\frac{1}{5} = 0.2$

7. $$\begin{array}{r} 0.3125 \\ 16\overline{)5.0000} \\ \underline{4\ 8\phantom{000}} \\ 20\phantom{00} \\ \underline{16\phantom{00}} \\ 40\phantom{0} \\ \underline{32\phantom{0}} \\ 80 \\ \underline{80} \\ 0 \end{array}$$

$\frac{5}{16} = 0.3125$

9. $$\begin{array}{r} 0.7 \\ 10\overline{)7.0} \\ \underline{7\ 0} \\ 0 \end{array}$$

$\frac{7}{10} = 0.7$

11. $$\begin{array}{r} 0.675 \\ 40\overline{)27.000} \\ \underline{24\ 0\phantom{00}} \\ 3\ 00\phantom{0} \\ \underline{2\ 80\phantom{0}} \\ 200 \\ \underline{200} \\ 0 \end{array}$$

$\frac{27}{40} = 0.675$

13. $$\begin{array}{r} 0.52 \\ 25\overline{)13.00} \\ \underline{12\ 5\phantom{0}} \\ 50 \\ \underline{50} \\ 0 \end{array}$$

$\frac{13}{25} = 0.52$

15. $$\begin{array}{r} 0.21875 \\ 32\overline{)7.00000} \\ \underline{6\ 4\phantom{0000}} \\ 60\phantom{000} \\ \underline{32\phantom{000}} \\ 280\phantom{00} \\ \underline{256\phantom{00}} \\ 240\phantom{0} \\ \underline{224\phantom{0}} \\ 160 \\ \underline{160} \\ 0 \end{array}$$

$\frac{7}{32} = 0.21875$

17. $$\begin{array}{r} 0.583 \\ 12\overline{)7.000} \\ \underline{60\phantom{00}} \\ 100\phantom{0} \\ \underline{96\phantom{0}} \\ 40 \\ \underline{36} \\ 4 \end{array}$$

$\frac{7}{12} = 0.583 = 0.58$ to nearest hundredth

19. $\frac{1}{18} = 0.0555\cdots = 0.0\bar{5}$

21. $\frac{3}{11} = 0.272727\cdots = 0.\overline{27}$

23. $7\frac{3}{4} = \frac{31}{4}$

$$\begin{array}{r} 7.75 \\ 4\overline{)31.00} \\ \underline{28} \\ 30 \\ \underline{28} \\ 20 \\ \underline{20} \\ 0 \end{array}$$

$7\frac{3}{4} = 7.75$

25. $\frac{3}{4} = 0.75 < 0.8$ so $\frac{3}{4} < 0.8$

27. $\frac{5}{16} = 0.3125 < 0.313$ so $\frac{5}{16} < 0.313$

**EXERCISES 9.5**

1. $0.9 = \frac{9}{10}$

3. $0.8 = \frac{8}{10} = \frac{4}{5}$

5. $0.37 = \frac{37}{100}$

7. $0.587 = \frac{587}{1000}$

9. $0.48 = \frac{48}{100} = \frac{12}{25}$

11. $0.58 = \frac{58}{100} = \frac{29}{50}$

13. $0.425 = \frac{425}{1000} = \frac{17}{40}$

15. $0.375 = \frac{375}{1000} = \frac{3}{8}$

17. $0.136 = \frac{136}{1000} = \frac{17}{125}$

19. $0.059 = \frac{59}{1000}$

21. $0.0625 = \frac{625}{10,000} = \frac{1}{16}$

23. $6.3 = 6\frac{3}{10}$

25. $2.17 = 2\frac{17}{100}$

27. $5.28 = 5\frac{28}{100} = 5\frac{7}{25}$

# CHAPTER 10
# RATIOS AND PROPORTION

**EXERCISES 10.1**

1. The ratio of 9 to 13: $\frac{9}{13}$

3. The ratio of 9 to 4: $\frac{9}{4}$

5. The ratio of 10 to 15: $\frac{10}{15} = \frac{2}{3}$

7. The ratio of 21 to 14: $\frac{21}{14} = \frac{3}{2}$

9. The ratio of 17 in. to 30 in.: $\frac{17 \text{ in.}}{30 \text{ in.}} = \frac{17}{30}$

11. The ratio of 12 mi to 18 mi: $\frac{12 \text{ mi}}{18 \text{ mi}} = \frac{12}{18} = \frac{2}{3}$

13. The ratio of 40 ft to 65 ft: $\frac{40 \text{ ft}}{65 \text{ ft}} = \frac{8}{13}$

15. The ratio of \$48 to \$42: $\frac{\$48}{\$42} = \frac{8}{7}$

17. The ratio of 75 sec to 3 min: $\frac{75 \text{ sec}}{3 \text{ min}} = \frac{75 \text{ sec}}{180 \text{ sec}} = \frac{5}{12}$

19. The ratio of 4 nickels to 5 dimes: $\frac{4 \text{ nickels}}{5 \text{ dimes}} = \frac{2 \text{ dimes}}{5 \text{ dimes}} = \frac{2}{5}$

21. The ratio of 2 days to 10 hours: $\frac{2 \text{ days}}{10 \text{ hours}} = \frac{48 \text{ hours}}{10 \text{ hours}} = \frac{24}{5}$

23. The ratio of 5 gal to 12 quarts: $\frac{5 \text{ gal}}{12 \text{ quarts}} = \frac{20 \text{ quarts}}{12 \text{ quarts}} = \frac{5}{3}$

25. The ratio of men to women: $\frac{7}{13}$; the ratio of women to men: $\frac{13}{7}$

27. Ratio of wins to games played: $\frac{9}{16}$; ratio of wins to losses: $\frac{9}{7}$

29. Ratio of yes votes to no votes: $\frac{4500}{3000} = \frac{3}{2}$

31. Ratio of cents to ounces: $\frac{192 \text{ cents}}{32 \text{ ounces}} = \frac{6 \text{ cents}}{1 \text{ ounce}}$

33. $\frac{6\frac{1}{2} \text{ oz}}{2\frac{3}{4} \text{ lb}} = \frac{6\frac{1}{2}}{2\frac{3}{4} \cdot 16} = \frac{\frac{13}{2}}{\frac{11}{4} \cdot \frac{16}{1}} = \frac{13}{2} \cdot \frac{4}{11 \cdot 16} = \frac{13}{88}$

35. $\frac{5\frac{3}{16} \text{ ft}}{2\frac{1}{8} \text{ yd}} = \frac{5\frac{3}{16}}{2\frac{1}{8} \cdot 3} = \frac{\frac{83}{16}}{\frac{17}{8} \cdot 3} = \frac{83}{16} \cdot \frac{8}{3 \cdot 17} = \frac{83}{102}$

37. $\frac{160 - 125}{125} = \frac{35}{125} = \frac{7}{25}$

39. $\frac{\$90 \text{ billion}}{\$450 \text{ billion}} = \frac{90}{450} = \frac{1}{5}$

41. $\frac{100 \text{ million bicycles}}{140 \text{ million cars}} = \frac{100 \text{ bicycles}}{140 \text{ cars}} = \frac{5 \text{ bicycles}}{7 \text{ cars}}$

## EXERCISES 10.2

1. $\frac{2}{3} = \frac{6}{9}$, means: 3, 6; extremes: 2, 9

3. $\frac{8}{11} = \frac{16}{22}$, means: 11, 16; extremes: 8, 22

5. $\frac{3}{8} = \frac{a}{32}$, means: 8, a; extremes: 3, 32

7. $\frac{x}{6} = \frac{5}{30}$, means: 6, 5; extremes: x, 30

9. $\frac{1}{4} = \frac{2}{7}$, means = 2 × 4 = 8; extremes = 1 × 7 = 7, false

11. $\frac{6}{7} = \frac{18}{21}$, means = 7 × 18 = 126; extremes = 6 × 21 = 126, true

13. $\frac{3}{5} = \frac{6}{10}$, means = 5 × 6 = 30; extremes = 3 × 10 = 30, true

15. $\frac{9}{10} = \frac{2}{7}$, means = 10 × 2 = 20; extremes = 9 × 7 = 63, false

17. $\frac{5}{8} = \frac{15}{24}$, means = 8 × 15 = 120; extremes = 5 × 24 = 120, true

19. $\frac{5}{12} = \frac{8}{20}$, means = 12 × 8 = 96; extremes = 5 × 20 = 100, false

21. $\frac{2}{5} = \frac{7}{9}$, means = 5 × 7 = 35; extremes = 2 × 9 = 18, false

23. $\frac{5}{8} = \frac{75}{120}$, means = 8 × 75 = 600; extremes = 5 × 120 = 600, true

25. $\frac{12}{7} = \frac{96}{50}$, means = 7 × 96 = 672; extremes = 12 × 50 = 600, false

27. $\frac{76}{24} = \frac{19}{6}$, means = 24 × 19 = 456; extremes = 76 × 6 = 456, true

29. $\frac{\frac{1}{2}}{4} = \frac{5}{40}$, means = 4 × 5 = 20; extremes = $\frac{1}{2}$ × 40 = 20, true

31. $\frac{\frac{2}{3}}{6} = \frac{1}{12}$, means = 6 × 1 = 6; extremes = $\frac{2}{3}$ × 12 = 8, false

33. $\frac{0.3}{4} = \frac{1}{20}$, means = 4 × 1 = 4; extremes = 0.3 × 20 = 6, false

35. $\frac{0.6}{0.12} = \frac{2}{0.4}$, means = 0.12 × 2 = 0.24; extremes = 0.6 × 0.4 = 0.24, true

37. (11.5)(3.6) = 41.4; (21.2)(14.5) = 307.4, false

39. (4.75)(2) = 9.5; (3.8)(2.5) = 9.5, true

41. (11.97)(24.86) = 297.5742; (61.34)(13.12) = 804.7808, false

EXERCISES 10.3

1. $\frac{x}{3} = \frac{6}{9}$
$9x = 3 \cdot 6$
$9x = 18$
$\frac{9x}{9} = \frac{18}{9}$
$x = 2$

3. $\frac{10}{n} = \frac{15}{6}$
$15n = 10 \cdot 6$
$15n = 60$
$\frac{15n}{15} = \frac{60}{15}$
$n = 4$

5. $\frac{4}{7} = \frac{y}{14}$
$7y = 4 \cdot 14$
$7y = 56$
$\frac{7y}{7} = \frac{56}{7}$
$y = 8$

7. $\frac{5}{8} = \frac{a}{16}$
$8a = 5 \cdot 16$
$8a = 80$
$\frac{8a}{8} = \frac{80}{8}$
$a = 10$

9. $\frac{8}{p} = \frac{6}{3}$
$6p = 8 \cdot 3$
$6p = 24$
$\frac{6p}{6} = \frac{24}{6}$
$p = 4$

11. $\frac{11}{a} = \frac{2}{44}$
$2a = 11 \cdot 44$
$2a = 484$
$\frac{2a}{2} = \frac{484}{2}$
$a = 242$

13. $\frac{35}{40} = \frac{7}{n}$
$35n = 40 \cdot 7$
$35n = 280$
$n = \frac{280}{35}$
$n = 8$

15. $\frac{a}{42} = \frac{5}{7}$
$7a = 42 \cdot 5$
$7a = 210$
$a = \frac{210}{7}$
$a = 30$

17. $\frac{18}{12} = \frac{12}{p}$
$18p = 12 \cdot 12$
$18p = 144$
$p = \frac{144}{18}$
$p = 8$

19. $\frac{x}{18} = \frac{64}{72}$
$72x = 18 \cdot 64$
$72x = 1152$
$x = \frac{1152}{72}$
$x = 16$

21. $\frac{6}{n} = \frac{75}{100}$
$75n = 6 \cdot 100$
$75n = 600$
$n = \frac{600}{75}$
$n = 8$

23. $\frac{5}{35} = \frac{a}{28}$
$35a = 5 \cdot 28$
$35a = 140$
$a = \frac{140}{35}$
$a = 4$

25. $\frac{12}{100} = \frac{3}{x}$

$12x = 3 \cdot 100$

$12x = 300$

$x = \frac{300}{12}$

$x = 25$

27. $\frac{p}{24} = \frac{25}{100}$

$120p = 24 \cdot 25$

$120p = 600$

$p = \frac{600}{120}$

$p = 5$

29. $\frac{\frac{1}{2}}{2} = \frac{3}{a}$

$\frac{1}{2}a = 2 \cdot 3$

$\frac{1}{2}a = 6$

$a = \frac{6}{\frac{1}{2}}$

$a = 6 \cdot \frac{2}{1}$

$a = 12$

31. $\frac{\frac{1}{4}}{12} = \frac{m}{96}$

$12m = \frac{1}{4} \cdot 96$

$12m = 24$

$m = \frac{24}{12}$

$m = 2$

33. $\frac{\frac{2}{5}}{8} = \frac{1.2}{n}$

$\frac{2}{5}n = 8 \cdot 1.2$

$\frac{2}{5}n = 9.6$

$n = \frac{9.6}{\frac{2}{5}}$

$n = 9.6 \cdot \frac{5}{2}$

$n = \frac{48}{2}$

$n = 24$

35. $\frac{0.2}{2} = \frac{1.2}{a}$

$0.2a = 2 \cdot 1.2$

$0.2a = 2.4$

$a = \frac{2.4}{0.2}$

$a = 12$

37. $\frac{p}{7} = \frac{8}{0.7}$

$0.7p = 7.8$

$0.7p = 56$

$p = \frac{56}{0.7}$

$p = 80$

39. $\frac{x}{3.3} = \frac{1.1}{6.6}$

$6.6x = 3.3 \cdot 1.1$

$6.6x = 3.63$

$x = \frac{3.63}{6.6}$

$x = 0.55$

41. $\frac{x}{2} = \frac{6}{4}$

$4x = 2 \cdot 6$

$4x = 12$

$x = \frac{12}{4}$

$x = 3$

43. $\frac{12}{8} = \frac{x}{4}$

$8x = 12 \cdot 4$

$8x = 48$

$x = \frac{48}{8}$

$x = 6$

## EXERCISES 10.4

1. $\frac{12 \text{ books}}{\$40} = \frac{18 \text{ books}}{x}$

$12x = 40 \cdot 18$

$12x = 720$

$x = \frac{720}{12}$

$x = 60$

18 books cost $60

3. $\frac{18 \text{ bags}}{90 \text{ cents}} = \frac{48 \text{ bags}}{x}$

$18x = 90 \cdot 48$

$18x = 4320$

$x = \frac{4320}{18}$

$x = 240$

48 bags cost 240 cents or $2.40.

5. $\frac{15 \text{ tapes}}{6 \text{ hours}} = \frac{x}{40 \text{ hours}}$

$6x = 15 \cdot 40$

$6x = 600$

$x = \frac{600}{6}$

$x = 100$

The workers can complete 100 players in 40 hours.

7. $\frac{3}{2} = \frac{2880}{x}$

$3x = 2880 \cdot 2$

$3x = 5760$

$x = \frac{5760}{3}$

$x = 1920$

There were 1920 no votes case.

9. $\frac{5}{6} = \frac{15}{x}$

$5x = 6 \cdot 15$

$5x = 90$

$x = \frac{90}{5}$

$x = 18$

The enlargement is 18 in. high.

11. $\frac{110}{5} = \frac{x}{12}$

$5x = 110 \cdot 12$

$5x = 1320$

$x = \frac{1320}{5}$

$x = 264$

Christy can travel 264 miles on 12 gallons.

13. $\frac{165}{3} = \frac{x}{8}$

$3x = 165 \cdot 8$

$3x = 1320$

$x = \frac{1320}{3}$

$x = 440$

In 8 hours the car will travel 440 mi.

15. $\frac{3}{7} = \frac{15}{x}$

$3x = 15 \cdot 7$

$3x = 105$

$x = \frac{105}{3}$

$x = 35$

The larger gear has 35 teeth.

17. $\frac{30}{500} = \frac{x}{1200}$

$500x = 30 \cdot 1200$

$500x = 36{,}000$

$x = \frac{36{,}000}{500}$

$x = 72$

72 defective parts should be expected in the shipment.

19. $\frac{212}{2} = \frac{x}{11}$

$2x = 11 \cdot 212$

$2x = 2332$

$x = \frac{2332}{2}$

$x = 1166$

The back should gain 1166 yards in the 11 game season.

21. $\frac{2\text{ lb}}{2500\text{ ft}^2} = \frac{x}{8750\text{ ft}^2}$

$2500x = 2 \cdot 8750$

$2500x = 17{,}500$

$x = \frac{17{,}500}{2500}$

$x = 7$

7 lbs of seed will be needed for 8750 $ft^2$.

23. $\frac{9}{15} = \frac{x}{40}$

$15x = 9 \cdot 40$

$15x = 360$

$x = \frac{360}{15}$

$x = 24$

The tree is 24 ft high.

25. $\frac{\frac{1}{2}}{50} = \frac{6}{x}$

$\frac{1}{2}x = 6 \cdot 50$

$\frac{1}{2}x = 300$

$x = \frac{300}{\frac{1}{2}}$

$x = 600$

The towns are 600 mi apart.

27. $\frac{\frac{5}{2}}{5000} = \frac{x}{7200}$

$5000x = \frac{5}{2} \cdot 7200$

$5000x = 18{,}000$

$x = \frac{18{,}000}{5000}$

$x = 3.6$

The car should burn 3.6 quarts in 7200 mi.

29. $\frac{10.5}{35} = \frac{15}{x}$

$10.5x = 35 \cdot 15$

$10.5x = 525$

$x = \frac{525}{10.5}$

$x = 50$

The 15 cm long piece of tubing has a mass of 50 gm.

31. $\frac{80}{5.2} = \frac{150}{x}$

$80x = 5.2 \cdot 150$

$80x = 780$

$x = \frac{780}{80}$

$x = 9.75$

The sales tax on the $150 item is $9.75.

33. $\frac{2}{6} = \frac{x}{72}$

$6x = 2 \cdot 72$

$6x = 144$

$x = \frac{144}{6}$

$x = 24$

The watch will gain 24 min in 3 days.

35. 5 quarts $= 5 \cdot 32$ ounces $= 160$ ounces

$\frac{4}{160} = \frac{1}{x}$

$4x = 160$

$x = \frac{160}{4}$

$x = 40$

To mix a batch of 1 cup of paste, 40 ounces of water should be used.

37. $\frac{7}{10} = \frac{x}{115{,}000}$

$10x = 7(115{,}000)$

$10x = 805{,}000$

$x = \frac{805{,}000}{10}$

$x = 80{,}500$ cars with 1 person

39. $\frac{7}{20} = \frac{x}{400}$

$20x = 7(400)$

$20x = 2800$

$x = 140$ million cars

# CHAPTER 11
# PERCENT

## EXERCISES 11.1

1. 35 of 100 squares are shaded: shaded portion is 35%

3. 3 of 4 parts are shaded: shaded portion is $\frac{3}{4} = 0.75 = 75\%$.

5. 53 of 100: 53%

7. 74 of 100: 74%

9. 3 of 10: $\frac{3}{10} = 0.3 = 30\%$

11. 27 of 50: $\frac{27}{50} = 0.54 = 54\%$

13. 23 of 50: $\frac{23}{50} = 0.46 = 46\%$

15. 5 of 20: $\frac{5}{20} = 0.25 = 25\%$

## EXERCISES 11.2

1. $6\% = \frac{6}{100} = \frac{3}{50}$

3. $75\% = \frac{75}{100} = \frac{3}{4}$

5. $65\% = \frac{65}{100} = \frac{13}{20}$

7. $50\% = \frac{50}{100} = \frac{1}{2}$

9. $46\% = \frac{46}{100} = \frac{23}{50}$

11. $66\% = \frac{66}{100} = \frac{33}{50}$

13. $150\% = \frac{150}{100} = \frac{3}{2} = 1\frac{1}{2}$

15. $166\frac{2}{3}\% = \frac{166\frac{2}{3}}{100} = \frac{\frac{500}{3}}{\frac{100}{1}}$

$= \frac{500}{3} \cdot \frac{1}{100} = \frac{500}{300}$

$= \frac{5}{3} = 1\frac{2}{3}$

17. $20\% = \frac{20}{100} = 0.2$

19. $35\% = \frac{35}{100} = 0.35$

21. $39\% = \frac{39}{100} = 0.39$

23. $5\% = \frac{5}{100} = 0.05$

25. $135\% = \frac{135}{100} = 1.35$

27. $240\% = \frac{240}{100} = 2.4$

29. $23.6\% = \frac{23.6}{100} = 0.236$

31. $6.4\% = \frac{6.4}{100} = 0.064$

33. $0.2\% = \frac{0.2}{100} = 0.002$

35. $7\frac{1}{2}\% = \frac{7.5}{100} = 0.075$

37. $85\% = \frac{85}{100} = \frac{17}{20}$

## EXERCISES 11.3

1. 0.08 : 0.08 = 8%

3. 0.05 : 0.05 = 5%

5. 0.18 : 0.18 = 18%

7. 0.86 : 0.86 = 86%

9. 0.4 : 0.40 = 40%

11. 0.7 : 0.70 = 70%

13. 1.10 : 1.10 = 110%

15. 4.40 : 4.40 = 440%

17. 0.065 : 0.065 = 6.5%

19. 0.025 : 0.025 = 2.5%

21. 0.002 : 0.002 = 0.2%

23. 0.004 : 0.004 = 0.4%

25. $\frac{1}{4} = 0.25$: 0.25 = 25%

27. $\frac{2}{5} = 0.40$ : 0.40 = 40%

29. $\frac{1}{5} = 0.2$ : 0.20 = 20%

31. $\frac{5}{8} = 0.625$ : 0.625 = 62.5%

33. $\frac{5}{16} = 0.3125$ : 0.3125 = 31.25%

35. $3\frac{1}{2} = 3.5$ : 3.50 = 350%

37. $\frac{1}{6} = \frac{1}{6} \cdot \frac{100}{100} = \frac{100}{6} \cdot \frac{1}{100}$
$= 16\frac{2}{3} \cdot \frac{1}{100}$
$= 16\frac{2}{3}\%$

39. $\frac{7}{9} = \frac{7}{9} \cdot \frac{100}{100} = \frac{7 \cdot 100}{9} \cdot \frac{1}{100}$
$= \frac{700}{9}\% = 77.777\cdots\%$
= 77.8% to nearest tenth of a percent

41. $\frac{(11.9 - 6.4)}{6.4} = \frac{5.5}{6.4} = 0.859375$
= 85.9% to nearest tenth percent

## EXERCISES 11.4

1. 23% of 400 is 92.
R% B A

3. 40% of 600 is 240.
R% B A

5. What is 7% of 325?
A R% B

7. 16% of what number is 56?
R% B A

9. 480 is 60% of what number?
A R% B

11. What percent of 120 is 40?
R% B A

13. rate: 5%, R%
base: 40,000, B
amount: what commission, A

15. rate: what percent, R%
base: 30 students, B
amount: 5, A

17. rate: 5.5%, R%
base: selling price, B
amount: $3.30, A

19. rate: 6%, R%
base: 9000 students, B
amount: how many, A

EXERCISES 11.5

1. $A = R\cdot B$
$A = (0.35)(600)$
$A = 210$

3. $R\cdot B = A$
$(0.45)(200) = A$
$90 = A$

5. $A = R\cdot B$
$A = (0.40)(2500)$
$A = 1000$

7. $R\cdot B = A$
$R\cdot 50 = 4$
$R = \frac{4}{50} = 0.8$
$R = 8\%$

9. $R\cdot B = A$
$R\cdot 500 = 45$
$R = \frac{45}{500} = 0.09$
$R = 9\%$

11. $R\cdot B = A$
$R\cdot 200 = 340$
$R = \frac{340}{200} = 1.7$
$R = 170\%$

13. $A = R\cdot B$
$46 = (0.08)B$
$\frac{46}{0.08} = B$
$575 = B$

15. $R\cdot B = A$
$(0.11)B = 55$
$B = \frac{55}{0.11}$
$B = 500$

17. $R\cdot B = A$
$(0.13)B = 58.5$
$B = \frac{58.5}{0.13}$
$B = 450$

19. $A = R\cdot B$
$A = (1.1)(800)$
$A = 880$

21. $A = R\cdot B$
$A = (1.08)(4000)$
$A = 4320$

23. $R\cdot B = A$
$R(120) = 210$
$R = \frac{210}{120} = 1.75$
$R = 175\%$

25. $R\cdot B = A$
$R(90) = 360$
$R = \frac{360}{90} = 4$
$R = 400\%$

27. $R\cdot B = A$
$(1.25)B = 625$
$B = \frac{625}{1.25}$
$B = 500$

29. $R\cdot B = A$
$1.1B = 935$
$B = \frac{935}{1.1}$
$B = 850$

31. $A = R\cdot B$
$A = (0.085)(300)$
$A = 25.5$

33. $A = R\cdot B$
$A = 0.1175(6000)$
$A = 705$

35. $A = R\cdot B$
$A = (0.0525)(3000)$
$A = 157.5$

37. $R\cdot B = A$
$R(800) = 60$
$R = \frac{60}{800} = 0.075$
$R = 7.5\%$

39. $R\cdot B = A$
$R\cdot 180 = 120$
$R = \frac{120}{180} = \frac{12}{18}\cdot 100\cdot \frac{1}{100}$
$R = \frac{1200}{18}\cdot \frac{1}{100}$
$R = 66\frac{2}{3}\%$

41. $R \cdot B = A$
$R(1200) = 750$
$R = \frac{750}{1200} = 0.625$
$R = 62.5\%$

43. $R \cdot B = A$
$(0.105)B = 420$
$B = \frac{420}{0.105}$
$B = 4000$

45. $R \cdot B = A$
$(0.13)B = 58.5$
$B = \frac{58.5}{0.13}$
$B = 450$

47. $R \cdot B = A$
$(0.075)B = 195$
$B = \frac{195}{0.075}$
$B = 2600$

49. 25.8% of 4000 : 25% of 4000 $= \frac{1}{4} \cdot 4000 = 1000$

51. 74.7% of 600 : 75% of 600 $= \frac{3}{4} \cdot 600 = 450$

53. 150% of 400 : 150% of 400 $= 1.5 \cdot 400 = 600$

## EXERCISES 11.6

1. $A = R \cdot B$
$A = 0.12(3400)$
$A = 408$
The interest on the loan is $408.

3. $A = R \cdot B$
$A = 0.26(550)$
$A = 143$
Robert has $143 withheld from his pay each week.

5. $R \cdot B = 1$
$R(2800) = 140$
$R = \frac{140}{2800} = 0.05$
$R = 5\%$
The commission rate is 5%.

7. $R \cdot B = A$
$R(1200) = 18$
$R = \frac{18}{1200} = 0.015$
$R = 1.5\%$
The monthly interest rate is 1.5%.

9. $R \cdot B = A$
$0.8B = 20$
$B = \frac{20}{0.8}$
$B = 25$
There were 25 questions on the test.

11. $R \cdot B = A$
$(0.105)B = 525$
$B = \frac{525}{0.105}$
$B = 5000$
Patty's loan was $5000.

13. $A = R \cdot B$
$A = 0.064(260)$
$A = 16.64$
The tax on a $260 purchase is $16.64.

15. $A = R \cdot B$
$A = 0.065(125,000)$
$A = 8125$
The salesperson's commission is $8125.

17. $R \cdot B = A$
$R \cdot 1200 = 102$
$R = \frac{102}{1200} = 0.085$
$R = 8.5\%$
The town's unemployment rate is 8.5%

19. $R \cdot B = A$
$R \cdot 60 = 15$
$R = \frac{15}{60} = 0.25$
$R = 25\%$
The dropout rate is 25%.

21. $R \cdot B = A$

$0.65B = 780$

$B = \frac{780}{0.65}$

$B = 1200$

1200 people responded to the survey.

23. $R \cdot B = A$

$(0.22)B = 209$

$B = \frac{209}{0.22}$

$B = 950$

Samuel's monthly salary is $950.

25. cost + markup = selling price

600 + 0.22(600) = selling price

600 + 132 = selling price

selling price = 732

The refrigerator will sell for $732.

27. ending value = initial value + increase in value

ending value = 26,000 + 0.25(26,000) = 26,000 + 6500

ending value = 32,500

The value of the lot at the end of the period was $32,500.

29. ending enrollment = beginning enrollment + increase in enrollment

$1064 = 950 + R(950)$

$114 = R(950)$

$R = \frac{114}{950} = 0.12 = 12\%$

The enrollment increased 12%.

31. sale price = regular price − discount

sale price = regular price − (discount rate)(regular price)

$382.50 = 450 - R \cdot 450$

$450R = 450 - 382.50$

$450R = 67.50$

$R = \frac{67.50}{450} = 0.15 = 15\%$

The discount rate is 15%.

33. (rate of increase)(original price) = increase

(0.14)(original price) = 2030

original price = $\frac{2030}{0.14}$ = 14,500

The price of the van before the increase was $14,500.

35. (rate of decrease)(number of employees) = (number of employees who left)

(0.055)(number of employees) = 66

number of employees = $\frac{66}{0.055}$

number of employees = 1200

In July of 1991 the company has 1200 employees.

37. value @ end = value @ beginning − decrease

value @ end = value @ beginning − (rate of decrease)(value @ beginning)

value @ end = 15,000 − 0.075(15,000) = 15,000 − 1125 = 13,875.

The stock was worth $13,875 @ the end of the six month period.

39. price = cost + markup = cost + (markup rate)(cost)

price = 11 + 0.25(11) = 11 + 2.75 = 13.75

The selling price is $13.75.

41. $4000 $\xrightarrow{06\%}$ $4240 $\xrightarrow{06\%}$ $4494.40

start　　year 1　　year 2

43. $4000 $\xrightarrow{05\%}$ $4200 $\xrightarrow{05\%}$ $4410 $\xrightarrow{05\%}$ $4630.50

start　　year 1　　year 2　　year 3

45. $\frac{145}{194.5} = 0.7455012853\cdots$

$= 74.6\%$ to nearest tenth percent

47. $A = R \cdot B$

$10.85 = 0.63B$

$B = \frac{10.85}{0.63}$

$B = 17.\bar{2}$ million barrels.

# CHAPTER 12

# THE ENGLISH SYSTEM OF MEASUREMENT

## EXERCISES 12.1

1. $8 \text{ ft} = 8 \text{ ft}\left(\frac{12 \text{ in.}}{\text{ft}}\right)$
   $8 \text{ ft} = 96 \text{ in.}$

3. $3 \text{ lb} = 3 \text{ lb}\left(\frac{16 \text{ oz}}{\text{lb}}\right)$
   $3 \text{ lb} = 48 \text{ oz}$

5. $360 \text{ min} = 360 \text{ min}\left(\frac{\text{hour}}{60 \text{ min}}\right)$
   $360 \text{ min} = 6 \text{ hours}$

7. $4 \text{ days} = 4 \text{ days}\left(\frac{24 \text{ hours}}{\text{day}}\right)$
   $4 \text{ days} = 96 \text{ hours}$

9. $16 \text{ qt} = 16 \text{ qt}\left(\frac{\text{gal}}{4 \text{ qt}}\right)$
   $16 \text{ qt} = 4 \text{ gal}$

11. $10{,}000 \text{ lb} = 10{,}000 \text{ lb}\left(\frac{\text{ton}}{2000 \text{ lb}}\right)$
    $10{,}000 \text{ lb} = 5 \text{ tons}$

13. $30 \text{ pt} = 30 \text{ pt}\left(\frac{\text{qt}}{2 \text{ pt}}\right)$
    $30 \text{ pt} = 15 \text{ qt}$

15. $64 \text{ oz} = 64 \text{ oz}\left(\frac{\text{lb}}{16 \text{ oz}}\right)$
    $64 \text{ oz} = 4 \text{ lb}$

17. $7 \text{ yd} = 7 \text{ yd}\left(\frac{3 \text{ ft}}{\text{yd}}\right)$
    $7 \text{ yd} = 21 \text{ ft}$

19. $39 \text{ ft} = 39 \text{ ft}\left(\frac{\text{yd}}{3 \text{ ft}}\right)$
    $39 \text{ ft} = 13 \text{ yd}$

21. $6 \text{ hours} = 6 \text{ hours}\left(\frac{60 \text{ min}}{\text{hour}}\right)$
    $6 \text{ hours} = 360 \text{ min}$

23. $80 \text{ fl oz} = 80 \text{ fl oz}\left(\frac{\text{pt}}{16 \text{ fl oz}}\right)$
    $80 \text{ fl oz} = 5 \text{ pt}$

25. $120 \text{ s} = 120 \text{ s}\left(\frac{\text{min}}{60 \text{ s}}\right)$
    $120 \text{ s} = 2 \text{ min}$

27. $7 \text{ gal} = 7 \text{ gal}\left(\frac{4 \text{ qt}}{\text{gal}}\right)$
    $7 \text{ gal} = 28 \text{ qt}$

29. $5 \text{ mi} = 5 \text{ mi}\left(\frac{5280 \text{ ft}}{\text{mi}}\right)$
    $5 \text{ mi} = 26{,}400 \text{ ft}$

31. $8 \text{ min} = 8 \text{ min}\left(\frac{60 \text{ s}}{\text{min}}\right)$
    $8 \text{ min} = 480 \text{ s}$

33. $192 \text{ hours} = 192 \text{ hours}\left(\frac{\text{day}}{24 \text{ hours}}\right)$
    $192 \text{ hours} = 8 \text{ days}$

35. $16 \text{ qt} = 16 \text{ qt}\left(\frac{2 \text{ pt}}{\text{qt}}\right)$
    $16 \text{ qt} = 32 \text{ pt}$

37. $7\frac{1}{4} \text{ hr} = 7\frac{1}{4} \text{ hr}\cdot\frac{60 \text{ min}}{\text{hr}}$
    $= 435 \text{ min}$

39. $56 \text{ oz} = 56 \text{ oz}\cdot\frac{\text{lb}}{16 \text{ oz}}$
    $= 3.5 \text{ lb}$

41. $225 \text{ s} = 225 \text{ s}\cdot\frac{\text{min}}{60 \text{ s}}$
    $= 3.75 \text{ min}$

43. $1.55 \text{ lb} = 1.55 \text{ lb}\cdot\frac{16 \text{ oz}}{\text{lb}}$
    $= 24.8 \text{ oz}$

45. $9 \text{ million tons} = 9 \text{ million tons}\cdot\frac{2000 \text{ lb}}{\text{ton}}$
    $= 18{,}000 \text{ million lb}$
    $= 18 \text{ billion lb}$

EXERCISES 12.2

1. $\begin{aligned} 4\text{ ft } 18\text{ in.} &= 4\text{ ft} + 12\text{ in.} + 6\text{ in.} \\ &= 4\text{ ft} + 1\text{ ft} + 6\text{ in.} \\ &= 5\text{ ft } 6\text{ in.} \end{aligned}$

3. $\begin{aligned} 7\text{ qt } 5\text{ pt} &= 7\text{ qt} + 4\text{ pt} + 1\text{ pt} \\ &= 7\text{ qt} + 2\text{ qt} + 1\text{ pt} \\ &= 9\text{ qt } 1\text{ pt} \end{aligned}$

5. $\begin{aligned} 5\text{ gal } 9\text{ qt} &= 5\text{ gal} + 8\text{ qt} + 1\text{ qt} \\ &= 5\text{ gal} + 2\text{ gal} + 1\text{ qt} \\ &= 7\text{ gal } 1\text{ qt} \end{aligned}$

7. $\begin{aligned} 9\text{ min } 75\text{ s} &= 9\text{ min} + 60\text{ s} + 15\text{ s} \\ &= 9\text{ min} + 1\text{ min} + 15\text{ s} \\ &= 10\text{ min } 15\text{ s} \end{aligned}$

9. $\begin{array}{r} 8\text{ lb } \ 7\text{ oz} \\ +\ 6\text{ lb } 15\text{ oz} \\ \hline 14\text{ lb } 22\text{ oz} \end{array}$ $\begin{aligned} &= 14\text{ lb} + 16\text{ oz} + 6\text{ oz} \\ &= 14\text{ lb} + 1\text{ lb} + 6\text{ oz} \\ &= 15\text{ lb} + 6\text{ oz} \end{aligned}$

11. $\begin{array}{r} 3\text{ hr } 20\text{ min} \\ 4\text{ hr } 25\text{ min} \\ +\ 5\text{ hr } 35\text{ min} \\ \hline 12\text{ hr } 80\text{ min} \end{array}$ $\begin{aligned} &= 12\text{ hr} + 60\text{ min} + 20\text{ min} \\ &= 12\text{ hr} + 1\text{ hr} + 20\text{ min} \\ &= 13\text{ hr} + 20\text{ min} \end{aligned}$

13. $\begin{aligned} 4\text{ lb } 7\text{ oz} + 3\text{ lb } 11\text{ oz} + 5\text{ lb } 8\text{ oz} &= 4\text{ lb} + 3\text{ lb} + 5\text{ lb} + 7\text{ oz} + 11\text{ oz} + 8\text{ oz} \\ &= 12\text{ lb} + 26\text{ oz} \\ &= 12\text{ lb} + 16\text{ oz} + 10\text{ oz} \\ &= 12\text{ lb} + 1\text{ lb} + 10\text{ oz} \\ &= 13\text{ lb } 10\text{ oz} \end{aligned}$

15. $\begin{array}{r} 9\text{ lb } 15\text{ oz} \\ -\ 5\text{ lb } \ 8\text{ oz} \\ \hline 4\text{ lb } \ 7\text{ oz} \end{array}$

17. $\begin{array}{r} 6\text{ hr } 30\text{ min} \\ -\ 3\text{ hr } 50\text{ min} \\ \hline \end{array} \longrightarrow \begin{array}{r} 5\text{ hr } 90\text{ min} \\ -\ 3\text{ hr } 50\text{ min} \\ \hline 2\text{ hr } 40\text{ min} \end{array}$

19. $\begin{array}{r} 5\text{ yd } 1\text{ ft} \\ -\ 2\text{ yd } 2\text{ ft} \\ \hline \end{array} \longrightarrow \begin{array}{r} 4\text{ yd } 4\text{ ft} \\ -\ 2\text{ yd } 2\text{ ft} \\ \hline 2\text{ yd } 2\text{ ft} \end{array}$

21. $\begin{aligned} 4 \times 13\text{ oz} &= 52\text{ oz} \\ &= 48\text{ oz} + 4\text{ oz} \\ &= 3(16\text{ oz}) + 4\text{ oz} \\ &= 3\text{ lb } 4\text{ oz} \end{aligned}$

23. $\begin{array}{r} 4\text{ ft } 5\text{ in.} \\ \times \qquad 3 \\ \hline 12\text{ ft } 15\text{ in.} \end{array}$ $\begin{aligned} &= 12\text{ ft} + 12\text{ in.} + 3\text{ in.} \\ &= 12\text{ ft} + 1\text{ ft} + 3\text{ in.} \\ &= 13\text{ ft } 3\text{ in.} \end{aligned}$

25. $\begin{aligned} \frac{4\text{ ft } 6\text{ in.}}{2} &= \frac{4\text{ ft}}{2} + \frac{6\text{ in.}}{2} \\ &= 2\text{ ft } 3\text{ in.} \end{aligned}$

27. $\begin{aligned} \frac{16\text{ min } 28\text{ s}}{4} &= \frac{16\text{ min}}{4} + \frac{28\text{ s}}{4} \\ &= 4\text{ min } 7\text{ s} \end{aligned}$

29. $\begin{array}{r} 4\text{ ft } \ 8\text{ in.} \\ 11\text{ ft } \ 7\text{ in.} \\ +\ 9\text{ ft } \ 3\text{ in.} \\ \hline 24\text{ ft } 18\text{ in.} \end{array}$ $\begin{aligned} &= 24\text{ ft} + 12\text{ in.} + 6\text{ in.} \\ &= 24\text{ ft} + 1\text{ ft} + 6\text{ in.} \\ &= 25\text{ ft } 6\text{ in.} \end{aligned}$

The total length of material needed is 25 ft 6 in.

31. $\begin{array}{r} 2\text{ yd} \\ -\ 2\text{ ft } 10\text{ in.} \\ \hline \end{array} \longrightarrow \begin{array}{r} 6\text{ ft} \\ -\ 2\text{ ft } 10\text{ in.} \\ \hline \end{array} \longrightarrow \begin{array}{r} 5\text{ ft } 12\text{ in.} \\ -\ 2\text{ ft } 10\text{ in.} \\ \hline 3\text{ ft } \ 2\text{ in.} \end{array}$

3 ft 2 in. of fabric remain.

33. (2 ft 6 in.) × 2 = 4 ft 12 in. = 4 ft + 12 in. = 4 ft + 1 ft
= 5 ft = 4 ft 12 in.

(1 ft 8 in.) × 2 = 2 ft 16 in. = 2 ft + 12 in. + 4 in. = 2 ft + 1 ft + 4 in.
= 3 ft 4 in.

$$\begin{array}{rr} & 4 \text{ ft} \quad 12 \text{ in.} \\ + & 3 \text{ ft} \quad 4 \text{ in.} \\ \hline & 7 \text{ ft} \quad 16 \text{ in.} \end{array} = 7 \text{ ft} + 12 \text{ in.} + 4 \text{ in.} = 7 \text{ ft} + 1 \text{ ft} + 4 \text{ in.} = 8 \text{ ft } 4 \text{ in.},$$

which is 8 in. less than 9 ft.

Yes, there will be enough for the frame.

35.
$$\begin{array}{rr} & 1 \text{ pt} \quad 9 \text{ fl oz} \\ + & 2 \text{ pt} \quad 10 \text{ fl oz} \\ \hline & 3 \text{ pt} \quad 19 \text{ fl oz} \end{array} = 3 \text{ pt} + 16 \text{ fl oz} + 3 \text{ fl oz} = 3 \text{ pt} + 1 \text{ pt} + 3 \text{ fl oz}$$
= 4 pt 3 fl oz

$$\begin{array}{rr} & 3 \text{ qt} \\ - & 4 \text{ pt } 3 \text{ fl oz} \\ \hline \end{array} \longrightarrow \begin{array}{rr} & 6 \text{ pt} \\ - & 4 \text{ pt } 3 \text{ fl oz} \\ \hline \end{array} \longrightarrow \begin{array}{rr} & 5 \text{ pt } 16 \text{ fl oz} \\ - & 4 \text{ pt } 3 \text{ fl oz} \\ \hline & 1 \text{ pt } 13 \text{ fl oz} \end{array}$$

1 pt 13 fl oz of the developer remain.

37. 6(2 lb 9 oz) = 12 lb 54 oz = 12 lb + 48 oz + 6 oz = 12 lb + 3 lb + 6 oz
= 15 lb 6 oz

The total weight of the packages to be mailed is 15 lb 6 oz.

39. 3(12 oz) = 36 oz; 2 lb 8oz = 2 lb + 8 oz = 32 oz + 8 oz
= 40 oz

The large can containing 2 lb 8 oz (= 40 oz) is the better buy.

41.
$$\begin{array}{rr} & 2 \text{ gal} \quad 3 \text{ qt} \quad 1 \text{ pt} \\ + & 3 \text{ gal} \quad 2 \text{ qt} \quad 1 \text{ pt} \\ \hline & 5 \text{ gal} \quad 5 \text{ qt} \quad 2 \text{ pt} \end{array} = 5 \text{ gal} + 4 \text{ qt} + 1 \text{ qt} + 1 \text{ qt}$$
= 5 gal + 1 gal + 2 qt
= 6 gal 2 qt

43.
$$\begin{array}{rr} & 13 \text{ yd } 15 \text{ ft } 10 \text{ in.} \\ - & 9 \text{ yd } 16 \text{ ft } 15 \text{ in.} \\ \hline \end{array} = \begin{array}{r} 12 \text{ yd } 17 \text{ ft } 22 \text{ in.} \\ 9 \text{ yd } 16 \text{ ft } 15 \text{ in.} \\ \hline 3 \text{ yd } \quad 1 \text{ ft } \quad 7 \text{ in.} \end{array}$$

45.
$$\begin{array}{rr} & 2 \text{ weeks} \quad 7 \text{ days} \quad 18 \text{ hr} \quad 40 \text{ min} \\ \times & 2 \\ \hline & 4 \text{ weeks} \quad 14 \text{ days} \quad 36 \text{ hr} \quad 80 \text{ min} \end{array} = 4 \text{ weeks} + 2 \text{ weeks} + 24 \text{ hr} + 12 \text{ hr} + 60 \text{ min} + 20 \text{ min}$$
= 6 weeks + 1 day + 12 hr + 1 hr + 20 min
= 6 weeks 1 day 13 hr 20 min

## EXERCISES 12.3

1. $P = 4 + 5 + 7 + 6 = 22$ ft

3. $P = 6 + 8 + 7 = 21$ yd

5. $P = 2(3) + 2(10) = 26$ in.

7. $P = 4(5 \text{ yd}) = 20$ yd

9. $C = \pi D = 3.14(5) = 15.7$
$C = 15.7$ ft

11. $P = 10\frac{1}{2} + 7\frac{1}{2} + 4\frac{1}{2}$
$P = 22\frac{1}{2}$ yd

13. $C = 2\pi R = 2(3.14)(3.75) = 23.55$
$C = 23.6$ ft to one decimal place

15. $C = \pi D = \left(\frac{22}{7}\right)\left(17\frac{1}{2}\right) = 55$
$C = 55$ in.

17. $P = 7 + \frac{1}{2}(\pi \cdot 9) + 7 + 9 = 7 + \frac{1}{2}(3.14)(9) + 7 + 9$

$P = 37.13$

$P = 37.1$ ft to one decimal place

19. $P = 7 + \frac{1}{2}(2\pi \cdot 4) + 7 + 8 = 7 + (3.14)(4) + 7 + 8 = 34.56$

$P = 34.6$ ft to one decimal place

21. cost = cost per ft × perimeter = $45[2(4) + 2(5)] = 45(8 + 10)$

cost = $45(18) = 810$ cents

It will cost \$8.10 to trim around the window.

23. 45 in. (top), 83 in. (left side), 83 in. (right side)

$83 + 45 + 83 = 211 \text{ in.} \cdot \frac{\text{ft}}{12 \text{ in.}} = \frac{211}{12} \text{ ft}$

cost = cost per ft × number of ft

cost = $60\left(\frac{211}{12}\right) = 1055$ cents

It will cost \$10.55 to weatherstrip the sides and top of door.

25. $54 \text{ in.} \frac{\text{ft}}{12 \text{ in.}} = \frac{54}{12} \text{ ft}$; $48 \text{ in.} \frac{\text{ft}}{12 \text{ in.}} = \frac{48}{12} \text{ ft} = 4 \text{ ft}$

$P = 2\ell + 2w = 2\left(\frac{54}{12}\right) + 2(4) = 17$ ft

Marion will need 17 ft of fringe.

27. distance = distance per lap × number of laps

= circumference × 3 = $\pi D \times 3$

= $(3.14)(1000)(3) = 9420$ yd

Robert jogs 9420 yd on his morning run.

## EXERCISES 12.4

1. $A = s^2 = 6^2$

$A = 36 \text{ yd}^2$

3. $A = \frac{1}{2}b \cdot h = \frac{1}{2}(8)(5)$

$A = 20 \text{ ft}^2$

5. $A = L \cdot W = (3.5)(1.5) = 5.25$

$A = 5.3 \text{ ft}^2$ to one decimal place

7. $A = bh = 7(5) = 35$

$A = 35 \text{ in.}^2$

9. $A = L \cdot W + \frac{1}{2}bh$

$A = 5 \cdot 4 + \frac{1}{2}(4)(2)$

$A = 24 \text{ ft}^2$

11. $A = \pi r^2$

$A = 3.14\ (7^2) = 153.86$

$A = 153.9 \text{ in.}^2$

13. $A = L \cdot W - s^2$

$A = 7 \cdot 5 - 2^2$

$A = 31 \text{ in.}^2$

15. $A = \pi r^2$

$A = 3.14(3.5)^2 = 38.465$

$A = 38.5 \text{ yd}^2$ to one decimal place

17. $A = L_1 \cdot W_1 + L_2 \cdot W_2$

$A = 15 \cdot 9 + 3 \cdot 6$

$A = 153 \text{ in.}^2$

19. $A = \pi r^2 = \frac{22}{7} \cdot \left[\frac{1}{2}\left(3\frac{1}{2}\right)\right]^2$

$A = 9\frac{5}{8} \text{ yd}^2$

21. $A = L \cdot W = 5.5(8)$; cost = cost per $\text{ft}^2 \cdot$ area

$A = 44 \text{ ft}^2$

cost = \$0.70(44)

cost = \$30.80

23. $A = \frac{1}{2}bh = \frac{1}{2}(30)(20)$; cost = cost per ft$^2$·area
$A = 300$ ft$^2$ cost = \$3·300
cost = \$900

25. $A = L \cdot W = 10(18)$; number of rolls = area ÷ area per roll
$A = 180$ ft$^2$ = 180 ÷ 90 = 2
2 rolls will be needed to insulate the ceiling of the room.

27. painted are = area of walls − area of windows − area of doors

$$= \underbrace{2(15\cdot 8) + 2(12\cdot 8)}_{\text{area of walls}} - \underbrace{(6)(4) - 4(3)}_{\text{area of windows}} - \underbrace{2(7)(3)}_{\text{area of doors}}$$

= 354 ft$^2$
The surface area of the walls that needs to be painted is 354 ft$^2$.

29. $A = \pi r^2 = (3.14)(2.5)^2$; cost = cost per ft$^2$·area
$A = 19.625$ = \$3(19.625) = \$58.875
It will cost \$58.88 to have the top refinished.

31. $A = \frac{1}{2}\pi r^2 = \frac{1}{2}(3.14)(9)^2 = 127.17$
$A = 127.2$ ft$^2$ to one decimal place

33. $A = s^2 = (110)^2 = 12100 \text{ yd}^2 \cdot \frac{\text{acre}}{4840 \text{ yd}^2} = 2.5$ acres
$A = 2.5$ acres

35. $A = L \cdot W = (18)(12)$; cost = cost per yd$^2$·area
$A = 216 \text{ ft}^2 \cdot \frac{\text{yd}^2}{9 \text{ ft}^2}$ cost = \$15·24
$A = 24$ yd$^2$ cost = \$360

37. $A_{\text{shaded}} = A_{\text{big square}} - A_{\text{small square}}$
$A_{\text{shaded}} = 6^2 - 2^2 = 36 - 4 = 32$
$A_{\text{shaded}} = 32$ in.$^2$

39. $A_{\text{shaded}} = A_{\text{square}} - A_{\text{circle}}$
$A_{\text{shaded}} = 20^2 - \pi(10)^2 = 20^2 - (3.14)(10)^2 = 86$
$A_{\text{shaded}} = 86$ ft$^2$

41. $A_{\text{shaded}} = A_{\text{rectangle}} + A_{\text{semi-circle}}$
$A_{\text{shaded}} = 6(5) + \frac{1}{2}(3.14)(3)^2 = 44.13$
$A_{\text{shaded}} = 44.1$ ft$^2$

43. $A = L \cdot W$
$= (15)(3)$
$= 45$ mi$^2$

45. $15 \text{ mi}^2 = 15 \text{ mi}^2 \cdot \frac{640 \text{ acres}}{\text{mi}^2}$
= 9600 acres

## EXERCISES 12.5

1. $V = L \cdot W \cdot H = 6 \cdot 6 \cdot 6$
$V = 216$ ft$^3$

3. $V = L \cdot W \cdot H = 6 \cdot 8 \cdot 2$
$V = 96$ in.$^3$

5. $V = L \cdot W \cdot H = (5.3)(1.7)(1.8) = 16.218$
$V = 16.2 \text{ in.}^3$

7. $V = L \cdot W \cdot H = (3)(2)(4.5)$
$V = 27 \text{ yd}^3$

9. $V = \pi r^2 h = (3.14)(2)^2(6) = 75.36$
$V = 75.4 \text{ in.}^3$

11. $V = \pi r^2 h = (3.14)(2)^2(1.8) = 22.608$
$V = 22.6 \text{ in.}^3$

13. $V = L \cdot W \cdot H = 5 \cdot 3 \cdot 2$
$V = 30 \text{ ft}^3$

15. $V = L \cdot W \cdot H = 9 \cdot 9 \cdot 6 = 486 \text{ ft}^3 \cdot \frac{\text{yd}^3}{27 \text{ yd}^3}$
$V = 18 \text{ yd}^2$

17. $V = \pi r^2 h = (3.14)(2)^2(5)$
$V = 62.8 \text{ in.}^3$

19. $W = 3 \text{ in.} \cdot \frac{\text{ft}}{12 \text{ in.}} = \frac{1}{4} \text{ ft} \cdot \frac{\text{yd}}{3 \text{ ft}} = \frac{1}{12} \text{ yd}$
$H = 9 \text{ ft} \frac{\text{yd}}{3 \text{ ft}} = 3 \text{ yd}$
$L = 48 \text{ ft} \frac{\text{yd}}{3 \text{ ft}} = 16 \text{ yd}$
$V = L \cdot W \cdot H = (16)\left(\frac{1}{12}\right)(3)$
$V = 4 \text{ yd}^3$

21. number of gallons = number of gallons per $\text{ft}^2 \cdot$ volume
number of gallons = $(7.5)(30)(20)(5)$
number of gallons = 22,500 gallons

23. $L = 36 \text{ ft} \cdot \frac{\text{yd}}{3 \text{ ft}} = 12 \text{ yd}$
$W = 15 \text{ ft} \cdot \frac{\text{yd}}{3 \text{ ft}} = 5 \text{ yd}$
$H = 3 \text{ in.} \cdot \frac{\text{yd}}{36 \text{ in.}} = \frac{1}{12} \text{ yd}$
$V = L \cdot W \cdot H = (12)(5)\left(\frac{1}{12}\right)$
$V = 5 \text{ yd}^3$

25. $V = \frac{1}{3}\pi r^2 h = \frac{1}{3}(3.14)(3)^2(10)$
$V = 94.2 \text{ in.}^3$

27. $V = \pi r^2 h$
$= (3.14)(5)^2(80)$
$= 6280 \text{ ft}^3$

# CHAPTER 13
# THE METRIC SYSTEM OF MEASUREMENT

EXERCISES 13.1

1. The height of a ceiling: 2.5 m
3. The height of a kitchen counter: 90 cm
5. The height of a two-story building: 7 m
7. The width of a roll of scotch tape: 12.7 mm
9. The thickness of a window glass: 5 mm
11. The length of a ballpoint pen: 16 cm
13. A playing card is 6 cm wide.
15. A doorway is 2 m high.
17. A basketball court is 28 m long.
19. The width of a nail file is 12 mm.
21. A recreation room is 6 m long.
23. A long-distance run is 35 km.
25. 3000 mm = 3 m
27. 8 m = 800 cm
29. $250 \text{ km} = 250 \text{ km}\,\frac{0.62 \text{ mi}}{1 \text{ km}} = \underline{155} \text{ mi}$
31. $9 \text{ cm} = 9 \text{ cm}\cdot\frac{0.394 \text{ in.}}{\text{cm}} = \underline{3.546} \text{ in.}$
33. 7000 m = 7 km
35. 8 cm = 80 mm
37. 5 km = 5000 m
39. 5 m = 5000 mm
41. $P = 2L + 2W = 2(4) + 2(2)$
    $P = 12$ cm
43. $P = 2L + 2W = 2(4) + 2(3)$
    $P = 14$ cm
45. $P = 4s = 4(25)$
    $P = 100$ mm
47. $C = 2\pi r = \pi D = 3.14(30)$
    $C = 94.2$ mm

EXERCISES 13.2

1. A nickel: 5 g
3. A flashlight battery: 80 g
5. A volkswagon rabbit: 1000 kg
7. A dinner fork: 50 g
9. A slice of bread: 20 g
11. A sugar cube: 2 g
13. A marshmallow's mass is 5 g.
15. A mg is $\frac{1}{1000}$ of a gram.
17. An electric razor's mass is 250 g.
19. A heavyweight boxer's mass is 98 kg.
21. A cigarette lighter's mass is 30 g.
23. A household broom's mass is 300 g.
25. 8 kg = 8000 g
27. $6 \text{ lb} = 6 \text{ lb}\cdot\frac{\text{kg}}{2.2 \text{ lb}} = 2.7 \text{ kg}$
29. $8 \text{ oz} = 8 \text{ oz}\cdot\frac{28 \text{ g}}{\text{oz}} = 224 \text{ g}$
31. 3 g = 3000 mg
33. $61.4 \text{ million tons} = 61.4 \text{ million tons}\cdot\frac{1000 \text{ kg}}{\text{ton}}$
    = 61400 million kg
    = 61.4 billion kg

EXERCISES 13.3

1. A bottle of wine has a volume of 750 mL.
3. A bottle of perfume has a volume of 15 mL.
5. A hot-water heater has a volume of 200 L.
7. A bottle of ink has a volume of 60 $cm^3$.
9. A jar of mustard has a volume of 150 mL.
11. A cream pitcher has a volume of 120 mL.

13. A can of tomato soup is 300 mL.

15. A saucepan holds 1.5 L.

17. A coffee pot holds 720 mL.

19. A car's engine capacity is 2000 $cm^3$. It is advertised as a 2.0 L model.

21. A mL is $\frac{1}{10}$ of a centiliter.

23. A garden sprinkler delivers 8 L of water per minute.

25. 7 L = 7000 mL

27. $4 \text{ qt} = 4 \text{ qt} \cdot \frac{0.95 \text{ L}}{\text{qt}} = 3.8 \text{ L}$

29. 8000 mL = 8 L

31. L = 5000 $cm^3$

33. 75 cL = 750 mL

35. 5 L = 500 cL

EXERCISES 13.4

1. A glass of iced tea: 4° C

3. A winter day in Fairbanks, Alaska: −15° C

5. The temperature in your classroom: 20° C

7. A hot summer day in Las Vegas: 45° C

9. Bathwater: 40° C

11. The temperature outside of an airliner at 30,000 ft: −25° C

13. 212° F = 100° C

15. $°F = \frac{9}{5}(°C) + 32 = \frac{9}{5}(40) + 32 = 104$ : 50° C = 104° F

17. $°C = \frac{5}{9}(°F - 32) = \frac{5}{9}(50 - 32) = 10$ : 50° F = 10° C

19. $°F = \frac{9}{5}(°C) + 32 = \frac{9}{5}(30) + 32 = 86$ : 30° C = 86° F

21. $°F = \frac{9}{5}(°C) + 32 = \frac{9}{5}(20) + 32 = 68$ : 20° C = 68° F

23. $°C = \frac{5}{9}(°F - 32) = \frac{5}{9}(104 - 32) = 40$ : 104° F = 40° C

25. $°F = \frac{9}{5}(°C) + 32 = \frac{9}{5}(15) + 32 = 59$ : 15° C = 59° F

27. $°C = \frac{5}{9}(°F - 32) = \frac{5}{9}(77 - 32) = 25$ : 77° F = 25° C

29. $°C = \frac{5}{9}(°F - 32) = \frac{5}{9}(535 - 32) = 279° C$; 535° F = 279° C

# CHAPTER 14
# GEOMETRY

## EXERCISES 14.1

1.

3.

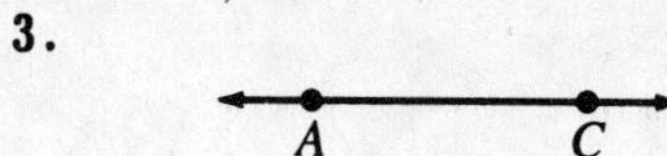


5. There are exactly two different rays that can be drawn through two points. True.

7. Two opposite sides of a square are parallel line segments. True.

9. ∡ ABC will always have the same measure as ∡ CAB. False

11.

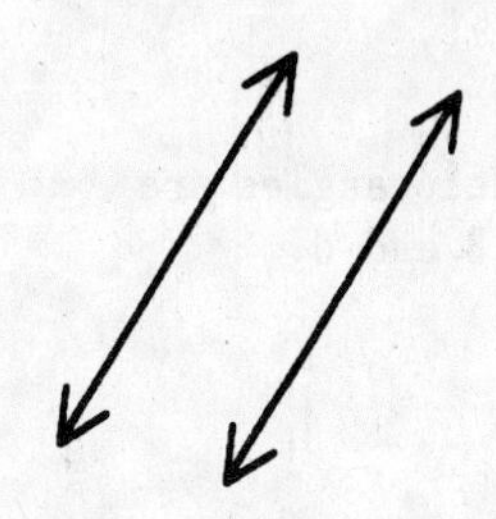

These two lines are parallel.

13.

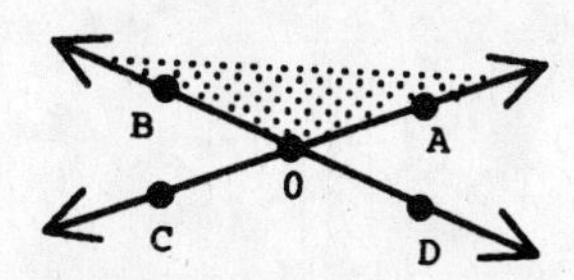


∡ BOA is highlighted

15.

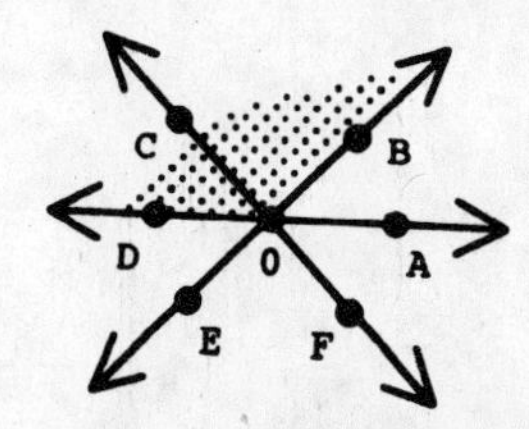


∡ DOB is highlighted

17.

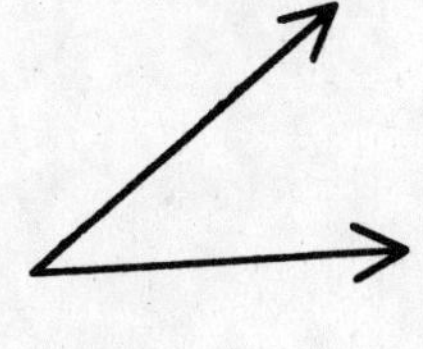

acute

19.

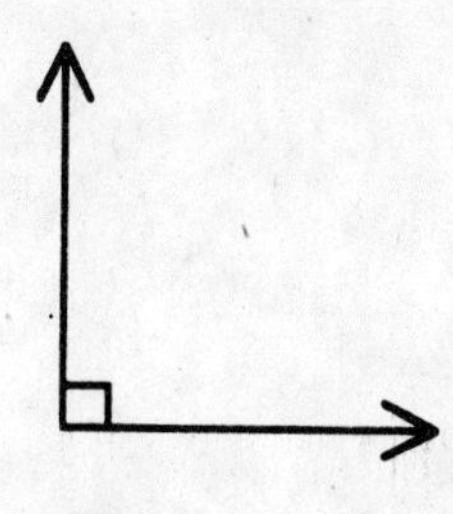

right

21.

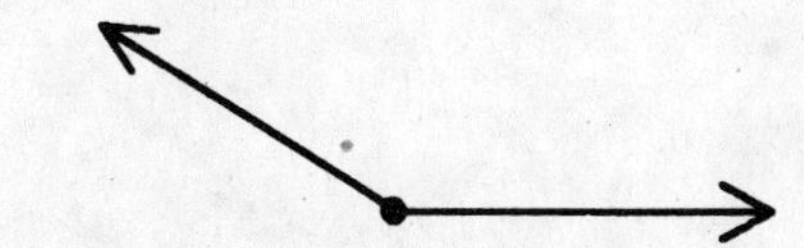

obtuse

23.

straight

25.

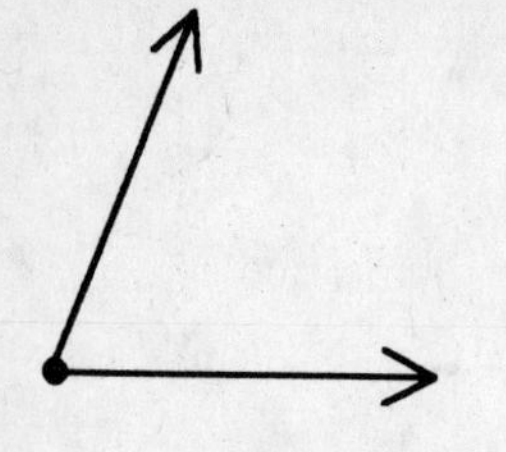

70°

27.

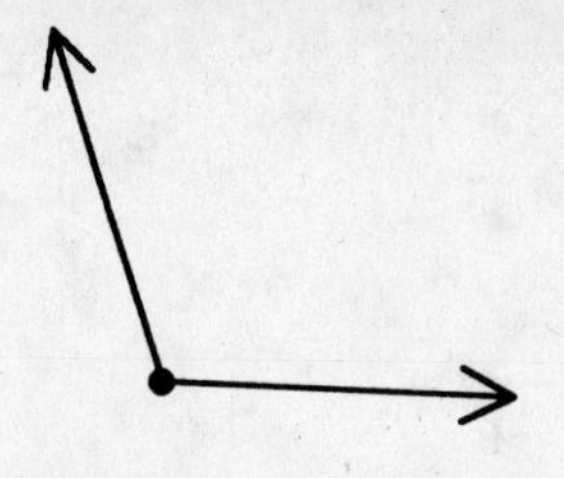

110°

29.

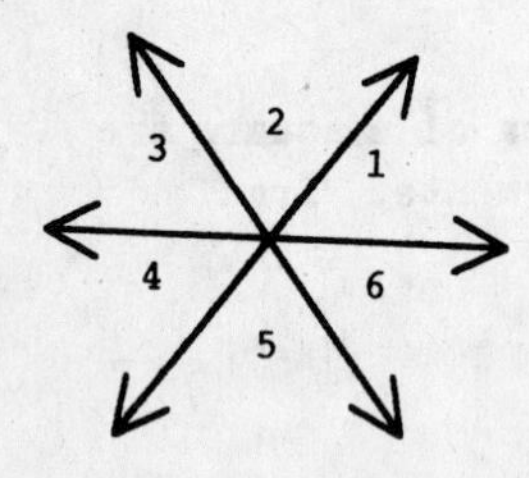


The pairs of verticle angles are
1 and 4, 2 and 5, 3 and 6.

EXERCISES 14.2

1.

acute

3.

acute

5.

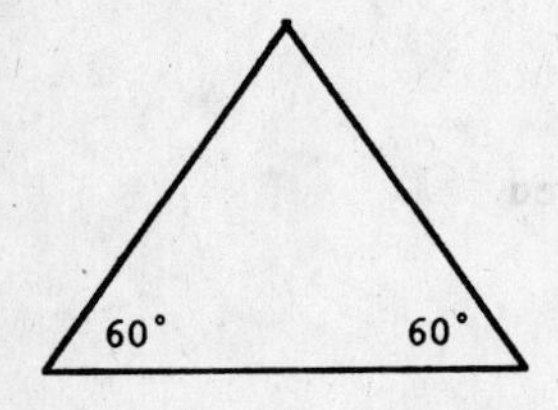


equililateral

7.

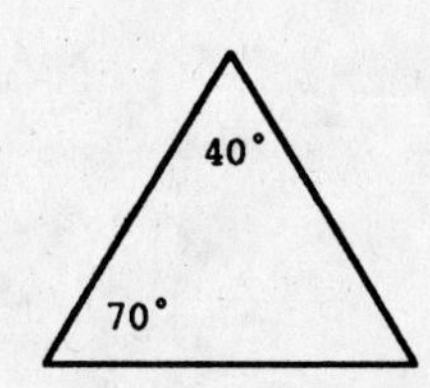


isosceles

9.

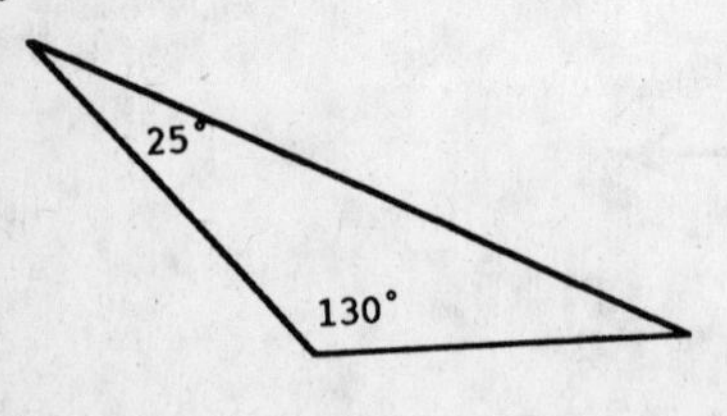


isosceles

11.

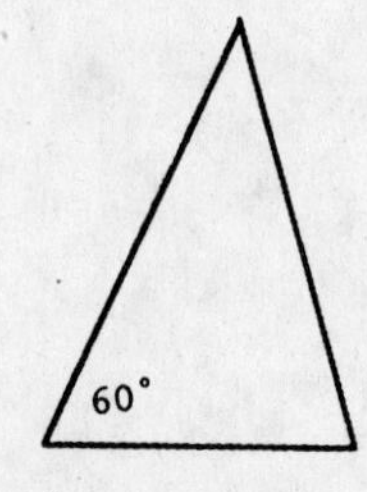


60°

13.

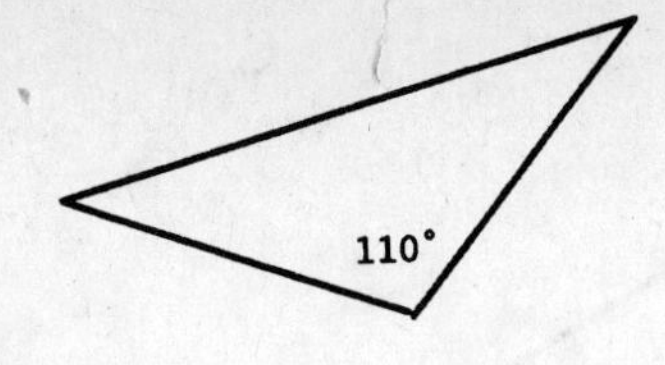


110°

15.

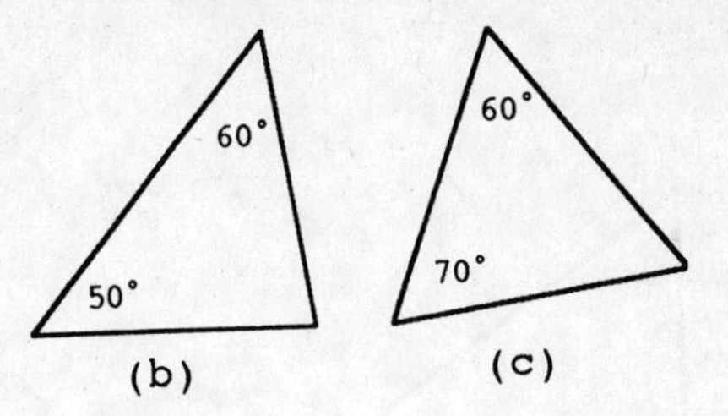


(b) and (c) are similar.

17.

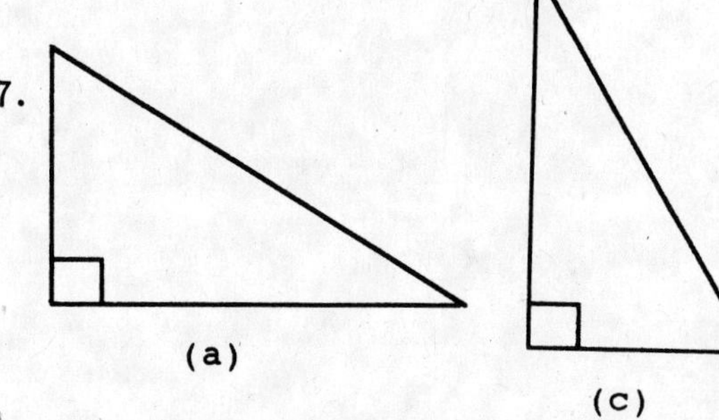


(a) and (c) are congruent.

19.

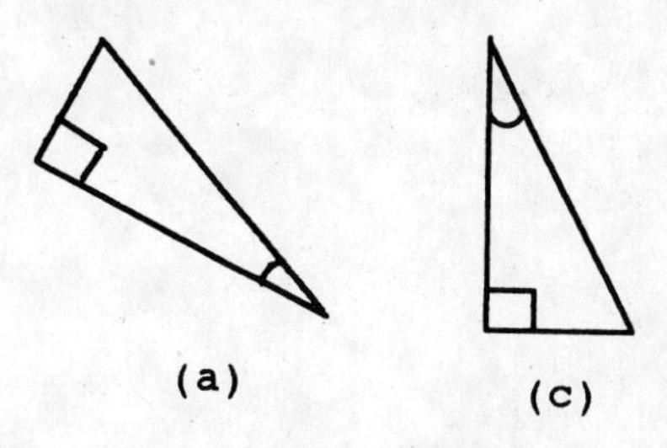


(a) and (c) are congruent.

EXERCISES 14.3

1. $\sqrt{64} = 8$

3. $\sqrt{169} = 13$

5.

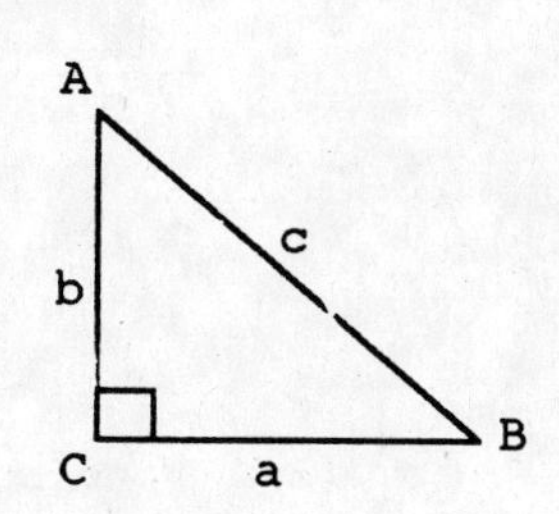


Side c is the hypotenuse of $\triangle ABC$.

7. $3^2 + 4^2 = 9 + 16 = 25 = 5^2$
3, 4, 5 is a perfect triple.

9. $7^2 + 12^2 = 49 + 144 = 193$
$\neq 13^2 = 169$
7, 12, 13 is not a perfect triple.

11. $8^2 + 15^2 = 64 + 225 = 289 = 17^2$
8, 15, 17 is a perfect triple.

13.

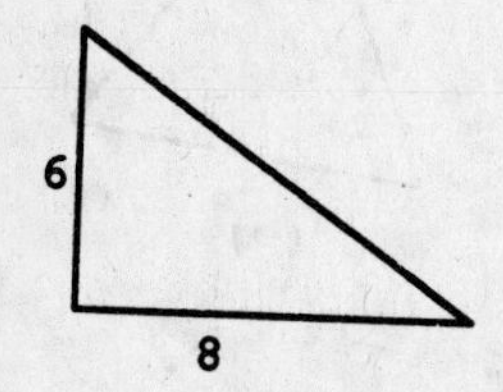


$x^2 = 6^2 + 8^2 = 36 + 64 = 100$

$x = 10$

15.

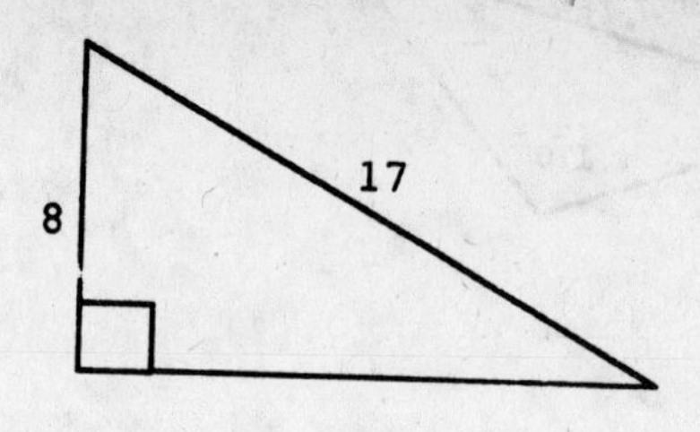


$8^2 + x^2 = 17^2$

$64 + x^2 = 289$

$x^2 = 225$

$x = 15$

# CHAPTER 15
# STATISTICS

EXERCISES 15.1

1. $\bar{x} = \frac{6 + 9 + 10 + 7 + 12}{5} = \frac{44}{5}$

   $\bar{x} = 8.8$

   There is no mode.

3. $\bar{x} = \frac{13 + 15 + 17 + 19 + 24 + 26}{6} = \frac{114}{6}$

   $\bar{x} = 19$

   There is no mode.

5. $\bar{x} = \frac{12 + 14 + 15 + 16 + 16 + 16 + 17 + 22 + 25 + 27}{10} = \frac{180}{10}$

   $\bar{x} = 18$

   $x_{mo} = 16$

7. $\bar{x} = \frac{9.1 + 9.2 + 9.6 + 9.5 + 9.7}{5} = \frac{47.1}{5}$

   $\bar{x} = 9.42$

   There is no mode.

9. $\bar{x} = \frac{11.16 + 11.25 + 11.32 + 11.44 + 11.75}{5} = \frac{56.92}{5}$

   $\bar{x} = 11.384$

   There is no mode.

11. 2, 3, 5, 6, 9

    $\tilde{x} = 5$

    $x_{mr} = \frac{2 + 9}{2} = \frac{11}{2}$

    $x_{mr} = 5.5$

13. 23, 24, 27, 31, 36, 38, 41

    $\tilde{x} = 31$

    $x_{mr} = \frac{23 + 41}{2} = \frac{64}{2}$

    $x_{mr} = 32$

15. 13, 25, 46, 47, 51, 68, 71

    $\tilde{x} = 47$

    $x_{mr} = \frac{13 + 71}{2} = \frac{84}{2}$

    $x_{mr} = 42$

17. 6.8, 7.0, 7.0, 7.1, 7.1, 7.4, 8.0, 8.5

    $x^1 = \frac{6.8 + 7.0 + 7.0 + 7.1 + 7.4 + 8.0 + 8.5}{8}$

    $= \frac{58.9}{8} = 7.3625$

    $\bar{x} = 7.4$ million metric tons of particulate matter

19. 6.8, 7.0, 7.0, 7.1, 7.1, 7.4, 8.0, 8.5

    $\tilde{x} = \frac{7.1 + 7.1}{2} = 7.1$

EXERCISES 15.2

1. 7 is the most common number of assignments completed.
3. 2 students have completed exactly 3 assignments.
5. (5 + 20 + 16) = 41 students have completed at least 6 assignments.
7. 32 percent of the students have completed exactly 8 assignments. $\frac{16}{50} = 0.32$
9. $(5 + 20 + 16) \div 50 = \frac{41}{50} = 0.82 = 82\%$ of the students have completed more than one-half the assignments.

11. 25 of the shoppers spent $10 or less.

13. (9 + 19) = 28 of the shoppers spend between $30.01 and $50.

15. (9 + 19 + 12 + 0 + 2 + 0 + 3 + 1) = 46 of the shoppers spent over $30.

## EXERCISES 15.3

1. The cost in 1987 was $700.

3. The decrease from 1986 to 1987 was $800 to $700 = $100.

5. The production in 1985 was 3 million cars.

7. 2, 4, 3, 6, 4, 4, 5 by year
   2, 3, 4, 4, 4, 5, 6 from low to high

   $\tilde{x} = 4$ million cars

9. production budgeted 45%

11. research budgeted 15%

13. miscellaneous budgeted 10%

15. 400 robberies occurred in November.

17. $\frac{400 + 500 + 300 + 300 + 400 + 600}{6} = \frac{2500}{6}$

$= 416.\overline{6}$

$\bar{x} = 417$

# CHAPTER 16
# THE INTEGERS

## EXERCISES 16.1

1. The opposite of 7 is −7, T.
3. −9 is an integer, T.
5. The opposite of −11 is 11, T.
7. $|-6| = -6$, F; absolute values are always positive.
9. −12 is not an integer, F.
11. $|7| = -7$, F; again, absolute values are always positive.
13. $-(-8) = 8$, T.
15. $|-20| = 20$, T.
17. $\frac{3}{5}$ is an integer, F.
19. 0.15 is not an integer, T.
21. $\frac{5}{7}$ is not an integer, T.
23. $-(-7) = -7$, F.
25. The absolute value of −10 is <u>10</u>.
27. $|-20| = \underline{20}$
29. The absolute value of 7 is <u>7</u>.
31. The opposite of 30 is <u>−30</u>.
33. $-(-6) = \underline{6}$.
35. $|50| = \underline{50}$
37. $-2 > -3$
39. $-20 < -10$
41. $|3| = 3$
43. $-4 < |-4|$

## EXERCISES 16.2

1. $-6 + (-5) = -11$
3. $8 + (-4) = 4$
5. $4 + (-6) = -2$
7. $7 + 9 = 16$
9. $(-11) + 5 = -6$
11. $-8 + (-7) = -15$
13. $(-16) + 15 = -1$
15. $-8 + 0 = -8$
17. $-9 + 10 = 1$
19. $-4 + 4 = 0$
21. $7 + (-13) = -6$
23. $-8 + 5 = -3$
25 $6 + (-6) = 0$
27. $(-10) + (-6) = -16$
29. $5\frac{3}{8} + \left(-3\frac{1}{8}\right) = 2\frac{2}{8} = 2\frac{1}{4}$
31. $-3.8 + 7.2 = 3.4$
33. $-4\frac{9}{16} + 7\frac{7}{16} = 7\frac{7}{16} + \left(-4\frac{9}{16}\right)$
$= 6\frac{23}{16} + \left(-4\frac{9}{16}\right) = 2\frac{14}{16} = 2\frac{7}{8}$
35. $-1.5 + (-0.3) = -1.8$
37. $4 + (-7) + (-5) = (-3) + (-5) = -8$
39. $-2 + (-6) + (-4) = -8 + (-4) = -12$
41. $-3 + (-7) + 5 + (-2) = -10 + 5 + (-2)$
$= -5 + (-2)$
$= -7$
43. $-7 + (-3) + (-4) + 8 = -10 + (-4) + 8$
$= -14 + 8$
$= -6$
45. $119 - 11 + 22 - 28 - 89 + 14 = 27$
27 million metric tons net export
47. $1 + 0 + 4 - 10 - 11 = -16$
−16 million metric tons

## EXERCISES 16.3

1. $5 - 7 = 5 + (-7) = -2$

3. $9 - 3 = 9 + (-3) = 6$

5. $-8 - 3 = -8 + (-3) = -11$

7. $-12 - 8 = -12 + (-8) = -20$

9. $-22 - 18 = -22 + (-18) = -40$

11. $3 - (-2) = 3 + (-(-2)) = 3 + (2) = 5$

13. $-2 - (-3) = -2 + (-(-3)) = -2 + (3) = 1$

15. $-5 - (-5) = -5 + (-(-5)) = -5 + (5) = 0$

17. $10 - (-5) = 10 + (-(-5)) = 10 + (5) = 15$

19. $38 - (-12) = 38 + (-(-12)) = 38 + (12) = 50$

21. $-15 - (-25) = -15 + (-(-25)) = -15 + (25) = 10$

23. $-25 - (-15) = -25 + (-(-15)) = -25 + (15) = -10$

25. $-0.5 - 1.5 = -0.5 + (-1.5) = -2$

27. $3.5 - (-2.5) = 3.5 + (-(-2.5)) = 3.5 + (2.5) = 6$

29. $5 - \left(-3\frac{1}{2}\right) = 5 + \left(-\left(-3\frac{1}{2}\right)\right) = 5 + \left(3\frac{1}{2}\right) = 8\frac{1}{2}$

31. $-2\frac{1}{4} - \left(-3\frac{3}{4}\right) = -2\frac{1}{4} + \left(-\left(-3\frac{3}{4}\right)\right) = -2\frac{1}{4} + 3\frac{3}{4} = 3\frac{3}{4} + \left(-2\frac{1}{4}\right) = 1\frac{2}{4} = 1\frac{1}{2}$

33. $-7 - (-5) - 6 = -7 + (-(-5)) - 6 = -7 + 5 - 6 = -2 - 6 = -2 + (-6) = -8$

35. $-10 - 8 - (-7) = -10 + (-8) - (-7) = -18 - (-7) = -18 + (-(-7)) = -18 + (7) = -11$

37. $22 - (-11) = 22 + (-(-11)) = 22 + (11) = 33$
The temperature dropped 33°F.

39. $34 + (12) + (-27) + (-6) + 15 = 46 + (-27) + (-6) + 15 = 19 + (-6) + 15 = 13 + 15 = 28$
Micki got off the elevator on the 28th floor.

41. $119 - 131 = -12$
decreased 12 million metric tons

43. $-28 - (-15) = -13$
declined 13 million metric tons

## EXERCISES 16.4

1. $7 \cdot 8 = 56$

3. $(4)(-3) = -12$

5. $(-8)(9) = -72$

7. $(-7)(-6) = 42$

9. $(-10)(0) = 0$

11. $(-8)(-8) = 64$

13. $(20)(-4) = -80$

15. $(-9)(-12) = 108$

17. $(-20)(1) = -20$

19. $(-40)(5) = -200$

21. $(1.8)(-0.2) = -0.36$

23. $\left(-\frac{7}{10}\right)\left(-\frac{5}{14}\right) = \frac{35}{140} = \frac{1}{4}$

25. $(-0.5)(-1.2) = 0.6$

27. $\left(\frac{5}{8}\right)\left(-\frac{4}{15}\right) = -\frac{20}{120} = -\frac{1}{6}$

29. $(-3)^2 = (-3)(-3) = 9$

31. $-3^2 = -(3)(3) = -9$

33. $(-4)^3 = (-4)(-4)(-4) = -64$

35. $(-2)^4 = (-2)(-2)(-2)(-2) = (4)(4) = 16$

37. $(-5)(3)(-8) = (-15)(-8) = 120$

39. $(-2)(-8)(-5) = (16)(-5) = -80$

41. $(2)(-5)(-3)(-5) = (-10)(-3)(-5)$
$= (30)(-5) = -150$

43. $(-4)(-3)(-6)(-2) = (12)(-6)(-2)$
$= (-72)(-2) = 144$

45. $C = \frac{5}{9}(F - 32)$

$= \frac{5}{9}(5 - 32)$ when $F = 5$.

$C = \frac{5}{9}(5 + (-32))$

$= \frac{5}{9}(-27)$

$C = -15$

A Celsius reading of $-15°$ C corresponds to $5°$ F.

EXERCISES 16.5

1. $15 \div (-3) = -5$

3. $\frac{48}{8} = 6$

5. $\frac{-50}{5} = -10$

7. $\frac{-24}{-3} = 8$

9. $-20 \div 5 = -4$

11. $-72 \div 8 = -9$

13. $\frac{60}{-15} = -4$

15. $18 \div (-1) = -18$

17. $\frac{0}{-9} = 0$

19. $-144 \div (-12) = 12$

21. $-7 \div 0$, undefined

23. $\frac{-150}{6} = -25$

25. $-4.5 \div (-0.9) = 5$

27. $-\frac{7}{9} \div \left(-\frac{14}{3}\right) = -\frac{7}{9}\left(-\frac{3}{14}\right) = \frac{1}{6}$

29. $\frac{7}{10} \div \left(-\frac{14}{25}\right) = \frac{7}{10}\left(-\frac{25}{14}\right) = -\frac{5}{4}$

31. $\frac{-7.5}{1.5} = -5$

33. $\frac{5 - 15}{2 + 3} = \frac{-10}{5} = -2$

35. $\frac{-6 + 18}{-2 - 4} = \frac{12}{-6} = -2$

37. $\frac{(5)(-12)}{(-3)(5)} = \frac{-60}{-15} = 4$

# CHAPTER 17

# ALGEBRAIC EXPRESSIONS AND EQUATIONS

## EXERCISES 17.1

1. When $a = 5$, $6a = 6(5) = 30$

3. When $b = 4$, $-2b = -2(4) = -8$

5. When $a = 5$, $b = 4$,
$5ab = 5(5)(4) = 100$

7. When $c = -2$, $d = -3$,
$6cd = 6(-2)(-3) = 36$

9. When $a = 5$, $b = 4$,
$2a + 3b = 2(5) + 3(4) = 10 + 12 = 22$

11. When $c = -2$, $d = -3$,
$3c + 4d = 3(-2) + 4(-3) = -6 - 12 = -18$

13. When $a = 5$, $c = -2$,
$3a - 4c = 3(5) - 4(-2) = 15 + 8 = 23$

15. When $c = -2$, $d = -3$
$5c - 2d = 5(-2) - 2(-3) = -10 + 6 = -4$

17. When $a = 5$, $a^2 = 5^2 = 25$

19. When $c = -2$, $c^3 = (-2)^3 = -8$

21. When $b = 4$, $2b^2 = 2(4)^2 = 2(16) = 32$

23. When $a = 5$, $d = -3$,
$a^2 + 2d^2 = 5^2 + 2(-3)^2 = 25 + 2(9)$
$= 25 + 18 = 43$

25. When $n = 4$, $t = 6$,
$2(n + t) = 2(4 + 6) = 2(10) = 20$

27. When $s = -3$, $n = 4$,
$3 = (s - n) = 3(-3 - 4) = 3(-7) = -21$

29. When $s = -3$, $t = 6$,
$4(s + t) = 4(-3 + 6) = 4(3) = 12$

31. When $n = 4$, $s = -3$,
$5(n + 3s) = 5(4 + 3(-3)) = 5(4 - 9)$
$= 5(-5) = -25$

33. When $m = -2$, $n = 4$, $t = 6$,
$m(2n + t) = -2(2(4) + 6) = -2(8 + 6)$
$= -2(14) = -28$

35. When $n = 4$, $t = 6$,
$2(n^2 + t^2) = 2(4^2 + 6^2) = 2(16 + 36)$
$= 2(52) = 104$

37. When $n = 4$, $s = -3$
$3(2n^2 + s^2) = 3(2(4)^2 + (-3)^2)$
$= 3(2(16) + 9) = 3(32 + 9)$
$= 3(41) = 123$

39. When $m = -2$, $n = 4$, $t = 6$,
$\frac{3mn}{t} = \frac{3(-2)(4)}{6} = \frac{-24}{6} = -4$

41. When $s = -3$, $t = 6$,
$\frac{5s + t}{s} = \frac{5(-3) + 6}{-3} = \frac{-15 + 6}{-3}$
$= \frac{-9}{-3} = 3$

43. When $m = -2$, $s = -3$, $n = 4$, $t = 6$,
$\frac{3m^2 - 4s}{3n - t} = \frac{3(-2)^2 - 4(-3)}{3(4) - 6} = \frac{3(4) + 12}{12 - 6}$
$= \frac{12 + 12}{6} = \frac{24}{6} = 4$

45. $d = rt = 55(4) = 220$

47. $h = 64t - 16t^2 = 64(2) - 16(2)^2$
$= 128 - 16(4) = 128 - 64 = 64$

49. $A = P(1 + rt) = 5000(1 + (0.08(2)))$
$= 5000(1 + 0.16)$
$= 5000(1.16)$
$= 5800$

## EXERCISES 17.2

1. $x - 3 = 4$
$\underline{\quad 3 = 3}$
$x = 7$
ck: $7 - 3 \stackrel{?}{=} 4$
$4 = 4$

3. $x + 4 = 10$
$\underline{-4 = -4}$
$x = 6$
ck: $6 + 4 \stackrel{?}{=} 10$
$10 = 10$

5. $x + 5 = -5$
$\underline{-5 = -5}$
$x = -10$
ck: $-10 + 5 \stackrel{?}{=} -5$
$-5 = -5$

7. $x - 6 = -5$
$\underline{\quad 6 = \ 6}$
$x = \ 1$

ck: $1 - 6 \stackrel{?}{=} -5$
$-5 = -5$

9. $x + 8 = -2$
$\underline{\quad -8 = -8}$
$x = -10$

ck: $-10 + 8 \stackrel{?}{=} -2$
$-2 = -2$

11. $\frac{7x}{7} = \frac{28}{7}$
$x = 4$

ck: $7(4) \stackrel{?}{=} 28$
$28 = 28$

13. $\frac{-10x}{-10} = \frac{-30}{-10}$
$x = 3$

ck: $-10(3) \stackrel{?}{=} -30$
$-30 = -30$

15. $\frac{6x}{6} = \frac{-42}{6}$
$x = -7$

ck: $6(-7) \stackrel{?}{=} -42$
$-42 = -42$

17. $\frac{-5x}{-5} = \frac{30}{-5}$
$x = -6$

ck: $-5(-6) \stackrel{?}{=} 30$
$30 = 30$

19. $4 \cdot \frac{x}{4} = 6 \cdot 4$
$x = 24$
ck: $\frac{24}{4} \stackrel{?}{=} 6$
$6 = 6$

21. $5 \cdot \frac{x}{5} = -10 \cdot 5$
$x = -50$
ck: $\frac{-50}{5} \stackrel{?}{=} -10$
$-10 = -10$

23. $2x + 5 = \ 9$
$\underline{\quad -5 = -5}$
$\frac{2x}{2} = \frac{4}{2}$
$x = 2$

ck: $2(2) + 5 \stackrel{?}{=} 9$
$4 + 5 = 9$
$9 = 9$

25. $4x - 5 = \ 7$
$\underline{\quad 5 = \ 5}$
$\frac{4x}{4} = \frac{12}{4}$
$x = 3$

ck: $4(3) - 5 \stackrel{?}{=} 7$
$12 - 5 \stackrel{?}{=} 7$
$7 = 7$

27. $3x - 10 = 17$
$\underline{\quad 10 = 10}$
$\frac{3x}{3} = \frac{27}{3}$
$x = 9$

ck: $3(9) - 10 \stackrel{?}{=} 17$
$27 - 10 \stackrel{?}{=} 17$
$17 = 17$

29. $4x - 3 = -11$
$\underline{\quad 3 = \ 3}$
$\frac{4x}{4} = \frac{-8}{4}$
$x = -2$

ck: $4(-2) - 3 \stackrel{?}{=} -11$
$-8 - 3 \stackrel{?}{=} -11$
$-11 = -11$

31. $5x + 6 = -14$
$\underline{\quad -6 = -\ 6}$
$\frac{5x}{5} = \frac{-20}{5}$
$x - -4$

ck: $5(-4) + 6 \stackrel{?}{=} -14$
$-20 + 6 \stackrel{?}{=} -14$
$-14 = -14$

33. $5 - 3x = -16$
$\underline{-5 \qquad = -\ 5}$
$\frac{-3x}{-3} = \frac{-21}{-3}$
$x = \ 7$

ck: $5 - 3(7) \stackrel{?}{=} -16$
$5 - 21 \stackrel{?}{=} -16$
$-16 = -16$

35. $6 - 5x = -9$
$\underline{-6 \qquad = -6}$
$\frac{-5x}{-5} = \frac{-15}{-5}$
$x = 3$

ck: $6 - 5(3) \stackrel{?}{=} -9$
$6 - 15 \stackrel{?}{=} -9$
$-9 = -9$

37. $\frac{x}{3} - 5 = \quad 3$
$\underline{\quad 5 = \quad 5}$
$3 \cdot \frac{x}{3} = 8 \cdot 3$
$x = 24$

ck: $\frac{24}{3} - 5 \stackrel{?}{=} 3$
$8 - 5 \stackrel{?}{=} 3$
$3 = 3$

39. $\frac{x}{5} + 7 = \quad 4$
$\underline{\quad -7 = \quad -7}$
$5 \cdot \frac{x}{5} = -3 \cdot 5$
$x = -15$

ck: $\frac{-15}{5} + 7 \stackrel{?}{=} 4$
$-3 + 7 \stackrel{?}{=} 4$
$4 = 4$

## EXERCISES 17.3

1. $7a + 5a = 12a$

3. $12b - 9b = 3b$

5. $5x - 5x = 0$

7. $-4y + 8y = 4y$

9. $-4m - 6m = -10m$

11. $7n - 9n = -2n$

13. $5a - 10a = -5a$

15. $-6b + 7b = b$

17.
$$\begin{aligned} 5x &= 4x + 3 \\ -4x &= -4x \\ \hline x &= 3 \end{aligned}$$
ck: $5(3) \stackrel{?}{=} 4(3) + 3$
$15 \stackrel{?}{=} 12 + 3$
$15 = 15$

19.
$$\begin{aligned} 7x &= 6x - 4 \\ -6x &= -6x \\ \hline x &= -4 \end{aligned}$$
ck: $7(-4) \stackrel{?}{=} 6(-4) - 4$
$-28 \stackrel{?}{=} -24 - 4$
$-28 = -28$

21.
$$\begin{aligned} 7x - 12 &= 5x \\ -5x + 12 &= -5x + 12 \\ \hline \frac{2x}{2} &= \frac{12}{2} \\ x &= 6 \end{aligned}$$
ck: $7(6) - 12 \stackrel{?}{=} 5(6)$
$42 - 12 \stackrel{?}{=} 30$
$30 = 30$

23.
$$\begin{aligned} 7x + 4 &= 6x - 3 \\ -6x - 4 &= -6x - 4 \\ \hline x &= -7 \end{aligned}$$
ck: $7(-7) + 4 \stackrel{?}{=} 6(-7) - 3$
$-49 + 4 \stackrel{?}{=} -42 - 3$
$-45 = -45$

25.
$$\begin{aligned} 3x - 5 &= 2x + 7 \\ -2x + 5 &= -2x + 5 \\ \hline x &= 12 \end{aligned}$$
ck: $3(12) - 5 \stackrel{?}{=} 2(12) + 7$
$36 - 5 \stackrel{?}{=} 24 + 7$
$31 = 31$

27.
$$\begin{aligned} 9x - 5 &= 8x - 7 \\ -8x + 5 &= -8x + 5 \\ \hline x &= -2 \end{aligned}$$
ck: $9(-2) - 5 \stackrel{?}{=} 8(-2) - 7$
$-18 - 5 \stackrel{?}{=} -16 - 7$
$-23 = -23$

29.
$$\begin{aligned} 6x + 3 &= 4x + 17 \\ -4x - 3 &= -4x - 3 \\ \hline \frac{2x}{2} &= \frac{14}{2} \\ x &= 7 \end{aligned}$$
ck: $6(7) + 3 \stackrel{?}{=} 4(7) + 17$
$42 + 3 \stackrel{?}{=} 28 + 17$
$45 = 45$

31.
$$\begin{aligned} 7x - 5 &= 4x + 10 \\ -4x + 5 &= -4x + 5 \\ \hline \frac{3x}{3} &= \frac{15}{3} \\ x &= 5 \end{aligned}$$
ck: $7(5) - 5 \stackrel{?}{=} 4(5) + 10$
$35 - 5 \stackrel{?}{=} 20 + 10$
$30 = 30$

33.
$$\begin{aligned} 9x - 3 &= 6x - 1 \\ -6x + 3 &= -6x + 3 \\ \hline \frac{3x}{3} &= \frac{2}{3} \\ x &= \frac{2}{3} \end{aligned}$$
ck: $9\left(\frac{2}{3}\right) - 3 \stackrel{?}{=} 6\left(\frac{2}{3}\right) - 1$
$6 - 3 \stackrel{?}{=} 4 - 1$
$3 = 3$

35.
$$\begin{aligned} 5x - 3 &= 6x - 10 \\ -5x + 10 &= -5x + 10 \\ \hline 7 &= x \end{aligned}$$
ck: $5(7) - 3 \stackrel{?}{=} 6(7) - 10$
$35 - 3 \stackrel{?}{=} 42 - 10$
$32 = 32$

37.
$$\begin{aligned} 5x + 3 &= 8x - 9 \\ -5x + 9 &= -5x + 9 \\ \hline \frac{12}{3} &= \frac{3x}{3} \\ 4 &= x \end{aligned}$$
ck: $5(4) + 3 \stackrel{?}{=} 8(4) - 9$
$20 + 3 \stackrel{?}{=} 32 - 9$
$23 = 23$

39.
$$\begin{aligned} 8x + 5 &= 12x + 2 \\ -8x - 2 &= -8x - 2 \\ \hline \frac{3}{4} &= \frac{4x}{4} \\ \frac{3}{4} &= x \end{aligned}$$
ck: $8\left(\frac{3}{4}\right) + 5 \stackrel{?}{=} 12\left(\frac{3}{4}\right) + 2$
$6 + 5 \stackrel{?}{=} 9 + 2$
$11 = 11$

41.
$$\begin{aligned} 4 - 3x &= 10 - 5x \\ -4 + 5x &= -4 + 5x \\ \hline \frac{2x}{2} &= \frac{6}{2} \\ x &= 3 \end{aligned}$$
ck: $4 - 3(3) \stackrel{?}{=} 10 - 5(3)$
$4 - 9 \stackrel{?}{=} 10 - 15$
$-5 = -5$

43.
$$\begin{aligned} 5 + 4x &= 17 + 7x \\ -17 - 4x &= -17 - 4x \\ \hline \frac{-12}{3} &= \frac{3x}{3} \\ -4 &= x \end{aligned}$$
ck: $5 + 4(-4) \stackrel{?}{=} 17 + 7(-4)$
$5 - 16 \stackrel{?}{=} 17 - 28$
$-11 = -11$

45.
$$\begin{aligned} 6 + 3x &= 14 - 9x \\ -6 + 9x &= -6 + 9x \\ \hline \frac{12x}{12} &= \frac{8}{12} \\ x &= \frac{8}{12} = \frac{2}{3} \end{aligned}$$
ck: $6 + 3\left(\frac{2}{3}\right) \stackrel{?}{=} 14 - 9\left(\frac{2}{3}\right)$
$6 + 2 \stackrel{?}{=} 14 - 6$
$8 = 8$

47. $$\begin{aligned} 9 - 3x &= 49 + 5x \\ -49 + 3x &= -49 + 3x \\ \hline \frac{-40}{8} &= \frac{8x}{8} \\ -5 &= x \end{aligned}$$

ck: $9 - 3(-5) \stackrel{?}{=} 49 + 5(-5)$

$9 + 15 \stackrel{?}{=} 49 - 25$

$24 = 24$

EXERCISES 17.4

1. The sum of r and s: $r + s$
3. w plus z: $w + z$
5. x increased by 2: $x + 2$
7. 10 more than y: $y + 10$
9. a minus b: $a - b$
11. x decreased by 9: $x - 9$
13. 6 less than r: $r - 6$
15. w times z: $wz$
17. The product of 5 and t: $5t$
19. The product of 9, p, and q: $9pq$
21. The sum of twice x and y: $2x + y$
23. x divided by 5: $\frac{x}{5}$
25. The quotient when a plus b is divided by 7: $\frac{a + b}{7}$
27. The difference of p and q divided by 4: $\frac{p - q}{4}$
29. The sum of a and 3, divided by the difference of a and 3: $\frac{a + 3}{a - 3}$
31. 7 more than a number: $x + 7$
33. 12 less than a number: $x - 12$
35. 8 times a number: $8x$
37. 5 more than 3 times a number: $3x + 5$
39. The quotient of a number and 9: $\frac{x}{9}$
41. The sum of a number and 7, divided by 6: $\frac{x + 7}{6}$
43. 8 more than a number divided by 8 less than that same number: $\frac{x + 8}{x - 8}$
45. (1) Find an unknown number.

    (2) Let x = unknown number

    (3) $2x + 7 = 35$

    (4) $2x = 28$
    $x = 14$

    (5) The unknown number is 14.

    $2(14) + 7 \stackrel{?}{=} 35$

    $28 + 7 \stackrel{?}{=} 35$

    $35 = 35$

47. (1) Find an unknown number.

(2) Let x = unknown number

(3) $5x - 12 = 78$

(4) $5x = 90$
$x = 18$

(5) The unknown number is 18.

$$5(18) - 12 \stackrel{?}{=} 78$$
$$90 - 12 \stackrel{?}{=} 78$$
$$78 = 78$$

49. (1) Find two unknown numbers.

(2) Let x = the smaller number
x + 5 = the larger number

(3) $x + (x + 5) = 55$

(4)
$$2x + 5 = 55$$
$$2x = 50$$
$$x = 25$$
$$x + 5 = 30$$

(5) The unknown numbers are 25 and 30.

$$25 + 30 \stackrel{?}{=} 55$$
$$55 = 55$$

51. (1) Find two unknown numbers.

(2) Let x = the smaller number
x + 7 = the larger number

(3) $x + (x + 7) = 33$

(4)
$$2x + 7 = 33$$
$$2x = 26$$
$$x = 13$$
$$x + 7 = 20$$

(5) The unknown numbers are 13 and 20.

$$13 + 20 \stackrel{?}{=} 33$$
$$33 = 33$$

53. (1) Find the number of votes received by two candidates in an election.

(2) Let x = the number of votes received by the losing candidate
x + 160 = the number of votes received by the winning candidate

(3) $x + (x + 160) = 3260$

(4)
$$2x + 160 = 3260$$
$$2x = 3100$$
$$x = 1550$$
$$x + 160 = 1710$$

(5) The losing candidate received 1550 votes, and the winning candidate received 1710 votes.

$$1550 + 1710 \stackrel{?}{=} 3260$$
$$3260 = 3260$$

55. (1) Find the cost of two appliances

(2) Let $x$ = cost of dryer
$x + 70$ = cost of washer

(3) $x + (x + 70) = 650$

(4)
$$\begin{aligned} 2x + 70 &= 650 \\ 2x &= 580 \\ x &= 290 \\ x + 70 &= 360 \end{aligned}$$

(5) The dryer cost \$290 and the washer cost \$360.

$$290 + 360 \stackrel{?}{=} 650$$
$$650 = 650$$

57. (1) Find the dimensions of a rectangle.

(2) Let $x$ = width of rectangle
$x + 5$ = length of rectangle

(3) $2x + 2(x + 5) = 98$

(4)
$$\begin{aligned} 2x + 2x + 10 &= 98 \\ 4x &= 88 \\ x &= 22 \\ x + 5 &= 27 \end{aligned}$$

x + 5

x

(5) The dimensions of the rectangle are 22 cm by 27 cm.

$$2(22) + 2(27) \stackrel{?}{=} 98$$
$$44 + 54 \stackrel{?}{=} 98$$
$$98 = 98$$

59. (1) Find the length and width of a rectangle.

(2) Let $x$ = length of rectangle
$x - 5$ = width of rectangle

(3) $2x + 2(x - 5) = 78$

(4)
$$\begin{aligned} 2x + 2x - 10 &= 78 \\ 4x &= 88 \\ x &= 22 \\ x - 5 &= 17 \end{aligned}$$

x

x − 5

(5) The length of the rectangle is 22 in. The width is 17 in.

$$2(22) + 2(17) \stackrel{?}{=} 78$$
$$44 + 34 \stackrel{?}{=} 78$$
$$78 = 78$$

61. (1) Find number of undernourished people in Africa and Asia.

(2) Let $x$ = number of undernourished people in Asia
$420 - x$ = number of undernourished people in Africa

(3) $2x = 420 - x$

(4) $2x + x = 420 - x + x$
$3x = 420$
$x = 140$ million undernourished people in Asia
$420 - x = 280$ million undernourished people in Africa

(5) Africa has twice as many undernourished people as Asia.

$2(140) \stackrel{?}{=} 280$
$280 = 280$

# NOTES

# NOTES

# NOTES

# NOTES

# NOTES

# NOTES

# NOTES